THE SCIENCE OF

ORPHAN BLACK

THE OFFICIAL COMPANION

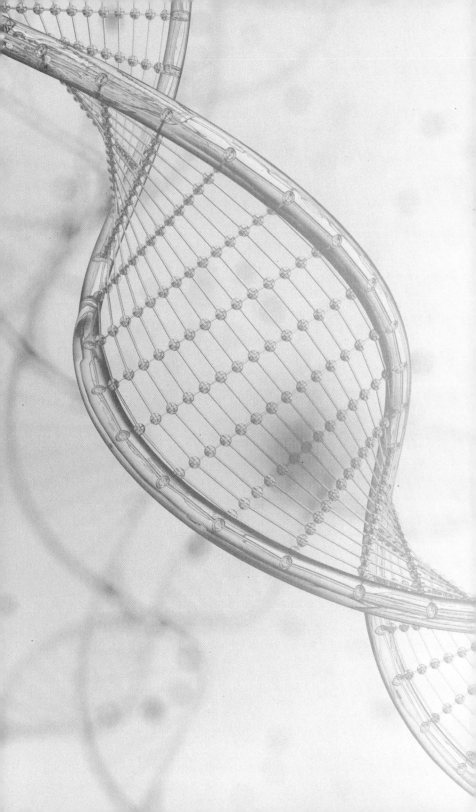

THE SCIENCE OF

ORPHAN BLACK

THE OFFICIAL COMPANION

CASEY GRIFFIN | NINA NESSETH

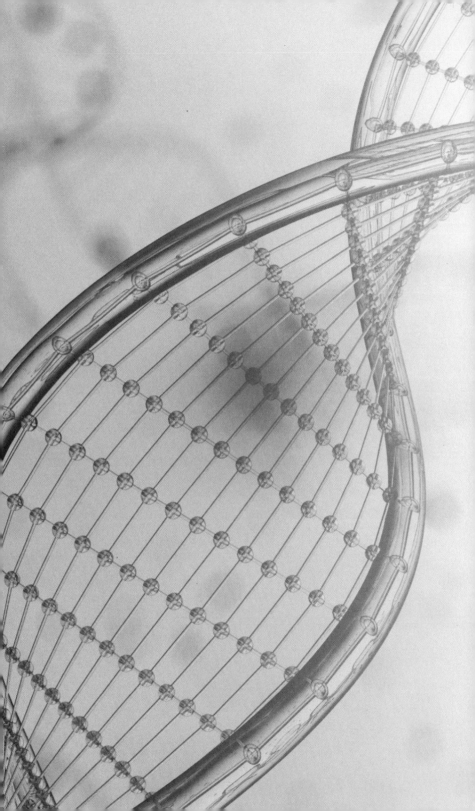

FOREWORD

by **COSIMA HERTER**

The entirety of the science you find in *Orphan Black* flowed from one original question which Graeme Manson put to me: *What do you know about clones?*

Clones.

This was the original guiding concept of Graeme Manson and John Fawcett's *Orphan Black* narrative, and the only science-related idea I had to work with in the beginning. In that first moment, late one summer evening over a glass of wine on his front porch, Graeme presented a fairly simple, but robust, idea he'd been thinking about: genetically identical people who look just like each other. But immediately this inspired in me more than just a vision of multiple carbon-copied reproductions of an original human genome all behaving as though they were the "same" person. *There are so many ways to think about clones*, I told him. *So many different kinds of organisms clone themselves — bees, potatoes, jellyfish, bacteria, water fleas, even single cells clone themselves — and for different reasons. But no one has successfully produced a human clone yet . . .*

Cloning, or what you'd call asexual reproduction, I explained, is a fascinating evolutionary strategy. Genetic clones are individuals, separated in space and time, and that alone prevents them from being *exactly* the same as one another. So, you also have to wonder if they'd have separate identities. Do you need to have a concept of "self" to have an identity? To have autonomy? To have agency? These were questions he'd have to consider, too, especially if he were thinking about *human* clones. There are so many ways to think about the biology of clones, and so many ways for the idea of clones to produce philosophical questions about identity and selfhood. *So, what exactly do you want to know about clones? Is this a biological question or an existential one?*

He meant both. If you met a clone of yourself, he wanted to know, would it necessarily have the same personality? Could you look exactly alike, be the same age — cloned like Dolly the sheep — and still have totally different identities? A whole cascade of questions followed. *Human clones would still be people, wouldn't they? How would human clones be made? Can they breed? Could they affect human evolution? How does evolution work? Could you produce a whole bunch of clones at once? Can we make synthetic people from clones?*

Genes alone do not determine who you are — they're only one component that contributes to your individual identity. Your genes alone are neither the fullness of your biology nor the extent of your "existential self" (as Graeme and I refer to it). Neither can be securely located in a string of DNA molecules. Developmental environment is one component. Socio-cultural environment another. Geography, climate, diet, pollutants, stress: these things are all contributors. Epigenetics are also important — phenomena that do not change the genome itself but affect how particular genes are expressed or silenced. To make a person, you need more ingredients than a genome can provide.

Genes do not determine your future, either. Your DNA is neither a scientifically accurate method of prediction nor a sacred divining rod. Your genes represent biological possibilities, some more probable than others. Genes do not oblige you to respond to the vicissitudes of life in any necessarily specific way. The chance events and contingent effects that construct our histories, the experiences and relationships we have, our mistakes and our successes, the knowledge we gain from them, along with biological compulsions, affect not only the choices we make but how we frame the range of choices we think we have. Genetic determinacy is a fantasy. Nature and nurture are not at odds; they function in tandem. For Graeme the storyteller, this was a gold mine of narrative and character possibilities. For me, this was a gorgeously rich opportunity to explore the complexities and controversies, the philosophies and politics embedded in and spinning out of reproductive technologies, synthetic biology, bioengineering, eugenics, theories of evolution, and the consequences of inserting our own agenda into biological processes.

From that one question, the scientific and philosophical issues built into the *Orphan Black* universe took shape. They were self-consciously and intentionally chosen, curated over several years of writing *Orphan Black*. While officially I held the title of science consultant, the choices were a collaborative effort. All the writers learned about, participated in, offered, debated, fought for, and accepted the science you read here in the following pages, and they wove a thrilling and compelling story from the pool of ideas we collected. Some ideas we chose very simply because we thought them elegant, provocative, perplexing, or ethically vexing. We mined historical events and now-defunct beliefs, as well as popular trends in contemporary research in biology, biotechnology and bioengineering, ethics, policy, law, and philosophy. Some we chose because they mobilized the narrative; some because we needed a way to solve a narrative problem. We tried to stay as true to the actual science as we could, but, of course, complete fidelity is mitigated by the medium in which the story is told. Science fiction is wonderful for problematizing our assumptions about humanity's place in the world, for speculating on future possibilities, for reframing past events in new ways.

Nina Nesseth and Casey Griffin have been writing about the science of *Orphan Black* for almost as long as the *OB* trip has been rolling along. I've followed their episodic analyses each season in *The Mary Sue* and was continually inspired, excited, flattered, but mostly humbled by their insight, their humor, and their ability to offer explanations and critiques in accessible, elegant, and straightforward language. It's remarkable how painstakingly they've identified, compiled, and explained the crazy science so beautifully. I couldn't be more excited to witness the birth of this book and the life and continuance of the conversation.

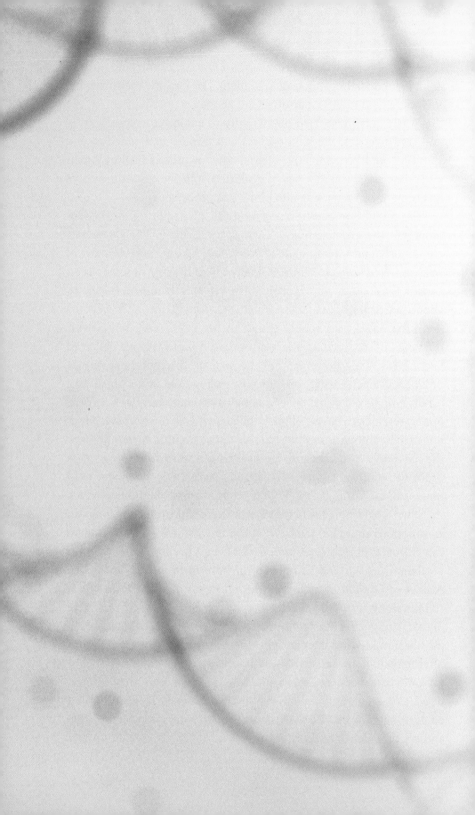

INTRODUCTION

WELCOME TO THE TRIP

NINA: If you're reading this right now, chances are you've already been inducted into Clone Club. If you're brand new to *Orphan Black*, then in the words of science nerd and Leda clone Cosima Niehaus: "*Welcome to the trip.*" People are drawn to *Orphan Black* for myriad reasons. But what makes *Orphan Black* extra-extra special for us is that every episode of its five-season run features science. Real science.

The thing about *Orphan Black*, unlike most science-fiction television out there, is that it doesn't take scientific concepts and stretch them thin until they are flimsy cellophane versions of real science. Most of the science explored on the show is real and follows current research interests. Notice that I'm saying *most* of the science: TV science is *never* perfect, mostly because the rules of the worlds that we see on TV are very seldom perfect mirrors of our own — especially when you squeeze those worlds into 42-minute episodes. Of course, that's where the fiction part of science fiction comes into play. The basic conceit of *Orphan Black* is this: in the 1970s, illegal human cloning experiments were launched. In the 1980s, the first human clones were born and were housed with surrogate families around the world, unaware of their clone status. In 2012, a number of the North American clones, now 28 years old, become self-aware and begin to seek answers and seek out each other.

Is this realistic? Well, as far as we know, a human clone has yet to be born. And if some human clone does secretly exist, it was certainly born more recently than the 1980s. The scientists at the Dyad Institute used more sophisticated technologies and techniques in the 1970s and '80s than actually existed at the time. But the techniques are those that we have historically used for real cloning projects. So, as far as suspending disbelief goes, it's simply a matter of accepting that, in the world of *Orphan Black*, some scientific breakthroughs may have been made just a few years earlier than in our reality. When all is said and done, *Orphan Black* remains one of the most true-to-science sci-fi shows out there. And if you're wondering where the show's science deviates toward the unreal, you're in luck! We'll highlight those moments for you, too.

CASEY: And the most exciting thing about *Orphan Black*, at least for scientists like us, is that the science on the show is never dumbed down or glossed over. The showrunners and writers know their audience is intelligent, and they treat us as such. What could be better for two scientists than a sci-fi show that discusses areas that we work in?! For me, a Ph.D. candidate in developmental and stem cell biology, and Nina, a professional science communicator, the science revelations on the show are exciting and thought-provoking; but for many of our non-scientist Clone Club friends, there's so much to explore in the science that has informed and shaped *Orphan Black*. We're here to help make this awesome show even more incredible by handing you your very own pair of lab goggles to examine the show through. Whether you're a casual science lover or you swore never to touch the subject again after your dreaded high-school chemistry class, this book will help you appreciate the subjects we love so much without scaring you away.

NINA: Allow us to be your geek monkey guides, if you will. Your all-access key card to the Dyad Institute. We've happily done the research for you (you're welcome), and we'll break the science down into (mostly) easy-to-digest terms. But be forewarned: this book contains spoilers for the show.

We've tried to answer the many questions we've fielded from Clone Club over the years, covering everything scientific to do with your favorite clones, from the history and science that led to real cloning projects, to an in-depth look at the Neolution technologies, to our observations about the mysterious clone disease. There are also some bonus materials if you want to dig a little deeper: a comparative timeline of landmark events in real science versus the science of *Orphan Black*. And if you've ever been unsure of what language Cosima and Scott are speaking whenever they science too hard (yep, that's science as a verb), we also have a handy glossary of terms. Hopefully, you'll find any answer you need within these pages (and if you don't, we encourage you to keep building your knowledge — that's what science is all about). We also have a special appearance by *Orphan Black* cocreator Graeme Manson and science consultant Cosima Herter for a conversation about science and their process of incorporating it into the show.

The last thing I'll mention is that you've probably noticed that there are two of us (I swear we're not clones), and that right now our voices are separate. You won't see that for the rest of the book. Just think of us as Team Science Megaforce.

CASEY: We are coming at you with maximum science fun! Because that's what this book is about: learning cool science facts in the context of an amazing show. And once you finish the book, you can refer to yourself as an official geek monkey, ready to take on the science in all aspects of the world around you (or as you continue to rewatch *Orphan Black* endlessly). The show may have ended, but the discussion lives on as real-world science continues to draw out connections to the science of the show. So, strap on your furry backward bike helmets and grab a clone phone, because the fun is about to begin!

CHAPTER 1

"HOW MANY OF US ARE THERE?"

THE HISTORY & SCIENCE OF CLONES

Imagine that you are Sarah Manning. You are standing on a train station platform late one night when something catches your eye. It's a woman and her back is turned to you. The woman takes off her suit jacket, folds it, and places it on the platform. She slips off her heels. She turns around.

The woman is identical to you. Wearing nicer clothes, granted, but her face? The spitting image.

Before you can get any answers, the woman has stepped out in front of a moving train. She's gone. Dead. You're an orphan and have never known your family. Your best guess — after stealing her purse and riffling through her things — is that this woman, Elizabeth Childs, must be your long-lost twin sister.

SARAH

What the hell, Fee? Did I have a
twin sister?

FELIX

Well, when you're a poor little
orphaned foster wretch, anything's
possible. Or so we tell ourselves.

(1.01 "Natural Selection")

It doesn't occur to Sarah Manning that "anything's possible" could include *clones*, and despite seeing Beth, and then meeting Katja, and then Alison and Cosima, the thought still doesn't occur to her. It takes a frustrated Alison blurting out that she, Cosima, and Sarah are all clones, all somebody's experiment, for the message to finally get across (although Cosima really wanted to float that whole clone thing a lot softer). But really, why should it have been obvious to Sarah that she was a clone? Human cloning is illegal, for one, and there is no record of a cloned human surviving past an embryonic stage, let alone into adulthood. The most famous clone — Dolly the sheep — was born in 1996. For Sarah, who is 28 years old in season one, to be a clone suggests that there was a much bigger cloning project happening more than a decade earlier in the 1980s. Is that unrealistic? Not really. Dolly may be a famous clone (for reasons we'll explain later), but she definitely wasn't the first.

First off, let's be clear about what we're talking about when we're talking about cloning. There's a whole world of possibilities for cloning out there other than just straight up creating identical copies of a being. We can clone on the tiniest scale and make copies of, say, fragments of DNA. This is known as molecular cloning. Then there's cellular cloning, which creates populations of identical cells from a single cell source. Cloned cells can be used for research (especially stem cell research) or to grow tissues. A really exciting aspect of cellular cloning is therapeutic cloning, which uses cloned cells for medical research, treatments, and transplants. And then, of course, there's organism cloning — known as reproductive cloning — by which we can create whole organisms that are identical to an original "parent" organism.

With *Orphan Black*, we're talking about reproductive cloning: creating complete living, breathing, genetically identical humans. We're talking about Sarah Manning. About Beth Childs. About Cosima Niehaus, and Alison Hendrix, and Rachel Duncan, and all of the clones that we've met over five seasons. We know that they are all (save for Charlotte Bowles) from a single generation of clones, born in the 1980s thanks to the efforts of professors Ethan and Susan Duncan and their research team at the Dyad Institute (including one religious extremist research assistant by the name of Henrik Johanssen). The cell line that gave rise to these clones came from a tissue sample from a woman named Kendall Malone, and a second tissue sample yielded clones for a second project, Project Castor, headed by the military.

WITH *ORPHAN BLACK*, WE'RE TALKING ABOUT REPRODUCTIVE CLONING: CREATING COMPLETE LIVING, BREATHING, GENETICALLY IDENTICAL HUMANS.

Ethan Duncan alludes to the ethical and biological challenges of cloning a human being, but he only scratches the surface of what a secret illegal international science experiment such as these would entail. In reality, the history of clones and cloning has been and continues to be filled with hurdles and controversies.

We meet the "father" of Project Leda, Ethan Duncan, in season two — a season full of examinations of scientists of different experiences, worldviews, moral backgrounds, and motivations — and as a way of weaving in the scientific themes of *Orphan Black*, Graeme Manson and Cosima Herter fittingly chose episode titles pulled from the works of Sir Francis Bacon. Sometimes referred to as the "Father of Experimental Philosophy," Francis Bacon was an English scholar in the late 16th and early 17th centuries who focused most of his work on philosophical and scientific ideas, most notably the development of the scientific method. The scientific method is a process of inquiry that involves developing a hypothesis based on observations and testing this hypothesis through experimentation. This, of course, ties into *Orphan Black* in that the clones came about as a result of experimentation based on laboratory observations and the hypothesis that human cloning was possible. Bacon's emphasis on inquiry and induction is mirrored by Ethan's work, making it a good match for the second season. Bacon also wrote many works focused on philosophy and religion, and particularly on morality. Once again, this dovetails perfectly with a season in which we meet the Proletheans and examine their questionable practices in the name of a higher power.

Interestingly, Bacon's *Novum Organum*, published in 1620, is a philosophical work discussing a new system of logic Bacon deemed superior to syllogism. We see syllogisms in *Orphan Black* during season three, as a means of testing the Castor clones for signs of "glitching." Bacon's work, specifically his focus on discussing ways to improve mankind, also mirrors themes brought up in season four. He touched on topics such as prolonging life, reforming law, and scientific innovation.

Aside from his academic writings, Bacon also wrote a novel, *New Atlantis*, which falls into the category of utopian-philosophical writings. While the plot is thin and the writing itself reads like an oven instruction manual, *New Atlantis* describes a location known as Salomon's House, a centrally organized research facility with the purpose of collecting data, conducting experiments, and using knowledge to improve the outside world. Sounds a bit like some of the centers on *Orphan Black*. Essentially, this novel set the stage for modern research centers and scientific communities, providing the blueprint for the likes of the British Royal Society, the National Institutes of Health, the Dyad Institute, and Revival.

HOW TO CREATE A CLONE

Cloning actually has botanical roots. The word "clone" is derived from the Greek κλών (*klon*), which translates to "twig." In 1903, Dr. Herbert J. Webber, a plant physiologist for the U.S. Department of Agriculture, adopted the word (he spelled it "clon") to describe the process of taking a graft, or cutting, from a plant and using it to propagate new plants asexually. To Webber, these cloned plants "are simply part of the same individual." Eventually, the term expanded

to include all forms of asexually produced life, but for a long time it was used only in the field of agriculture.

Let's fast-forward to 1952, some 45 years before Dolly was introduced to the world and close to 30 years before Dyad was working on Project Leda. In that year, two scientists, Robert Briggs and Thomas King, cloned frogs from tadpole cells, using a low-tech cloning technique known as embryo twinning, which basically mimics how identical twins are formed in nature. Very early in development, an embryo splits in two, and each half of the embryo develops separately to eventually create two individual, but genetically identical, beings. To clone the frogs, Briggs and King took early tadpole embryos, physically separated individual cells, and allowed them to grow in a petri dish. Since the embryonic cells came from the same fertilized egg, the tadpoles were genetically identical.

About a decade later, the experiment was repeated by another scientist, John Gurdon, who successfully cloned frogs from slightly older tadpoles. This may not sound like such a big deal, but it was: the reason why Briggs and King had used very early tadpole embryos was because, at that point, the cells are undifferentiated. Their functions — whether they will form the frog's skin or muscles or eyeballs — haven't been triggered yet. These cells have the potential to become any tissue. Stem cells are a class of undifferentiated cells. Once cells mature, they begin to specialize and find their places as skin, muscle, or eyeball cells. Scientists in the 1960s figured that once these cells had specialized, there was no going back to an undifferentiated state.

Let's skip ahead a few years to 1993. At this point in the world of *Orphan Black*, the Leda clones would be around nine years old. But in the real world, this is when Drs. Robert Stillman and Jerry Hall cloned human embryos at George Washington University. Yep, you read that right: you probably haven't heard about it before, but human embryos have been cloned. Stillman and Hall took 17 human embryos and separated them into 48 embryos using pretty much the same methods as scientists had with animal embryos. None of these embryos lived very long, but that's exactly what Stillman and Hall intended. They weren't trying to make human clones that would live into adulthood. In fact, they were both vocally *against* making adult clones and specifically worked with abnormal embryos that were unlikely to survive. They were trying to develop strategies for improving their *in vitro* fertilization (IVF) program. Today, *in vitro* (which translates to "in glass") technologies allow for scientists to fertilize egg cells with sperm *outside* of the human body (in a glass test tube, for example, or culture dish) to assist people who have difficulties conceiving. Henrik Johanssen used *in vitro* technology to harvest Helena's ova, fertilize them with his own sperm, and then implant the embryos (Helena's "babies") into both Helena and his daughter Gracie. For Stillman and Hall, more embryos strictly meant more opportunities for implantation for patients to conceive a child. Obviously, when you consider the potential for other uses, especially in the wrong hands — we're looking at you, Proletheans — this entire project raised some major ethical red flags. In 1994, it was found that Stillman and Hall had never obtained approval from George Washington University's review board for their project. After an investigation, they were instructed to destroy their data.

And then a few years later, Dolly, a clone of a six-year-old Finn-Dorset sheep, was born in Scotland on July 5, 1996, though her birth wasn't announced until February 1997. Obviously, Dolly was not the first cloned animal around, so that's not what caused so much fanfare. Rather, the excitement was because she was cloned from mammary cells from the udder of an adult sheep. Cells that had already differentiated. Something that was thought to be impossible. The scientists chose mammary cells specifically because they develop very quickly during pregnancy. (Dolly's name is actually a reference to Dolly Parton, a bit of a joke about the cells' origins. Her original name, though, was 6LL3.) Finn-Dorsets are completely white sheep; in order to make it immediately clear that Dolly was not related to the surrogate ewe that would be carrying her to term, the Finn-Dorset nucleus was implanted into the egg cell of a Scottish Blackface (which, if you couldn't tell by the name, is a black-faced sheep), and the surrogate ewe was another Scottish Blackface. It's worth mentioning that the sheep that Dolly was cloned from was not a living sheep. The udder cell that was manipulated to make Dolly was one among many that had been kept frozen in a vial for a good three years. That's why there aren't any family photos of Dolly with her "original," but there are some floating around of her with her surrogate!

Dolly was cloned using a technique called somatic cell nuclear transfer (SCNT, for short). In season three's "Newer Elements of Our Defense," Sarah learns that Henrik Johanssen previously worked with Dyad on Project Leda and continued his work after the laboratory at Dyad was destroyed. His uncovered experiment notes reveal that SCNT was the process likely used to create Sarah Manning and the other clones. SCNT works by taking a somatic cell (any cell that isn't an egg or sperm) from an organism and removing its nucleus, which contains the cell's DNA. The rest of the cell is discarded and the nucleus is implanted into an egg cell that has had its nucleus removed. So how do you remove a nucleus? First, you hold the cell steady by creating light suction with a pipette. Then, you use a tiny glass needle to remove a piece of the zona pellucida, a protective membrane that surrounds the egg cell. You reinsert the needle and use it to extract the nucleus and any polar bodies (which would also contain genetic material). *Voilà*, the egg cell (now called an enucleated ovum) is ready for a new nucleus. A small pulse of electricity helps the somatic nucleus incorporate into the egg cell and triggers cell division. After so many divisions, it becomes an early stage embryo, known as a blastocyst, with about 100 or so cells. The DNA within those cells should be identical (barring any random mutations — because they're gonna happen) to the DNA from the original organism. This embryo can then be implanted into a host uterus to grow.

HELENA

BAAAAAA.

(1.07 "Parts Developed in an Unusual Manner")

Orphan Black makes numerous nods to Dolly in the show: in season one, Helena makes repeated reference to the clones being sheep thanks to her mentor Tomas's brainwashing tactics. In "Parts Developed in an Unusual Manner," Tomas directs Helena to kill her clones by dehumanizing them: "The path to the shepherd is through the sheep." (And in this same episode, Kira is heard playing the melody for "Baa, Baa, Black Sheep" for Sarah on the piano.) In season four, a new clone known initially as M.K. protects her identity by wearing a novelty sheep mask.

COSIMA

[...] a clone in a mask. What's up with that?

SARAH

Art just said she wore a sheep mask. What?

COSIMA

(laughing)

Dolly the sheep.

ALISON

For the record, I do not find that even remotely amusing.

(4.02 "Transgressive Border Crossing")

Dolly wasn't the only clone to make her debut in 1997: on October 3 of that year, Cumulina the cloned mouse was born at the University of Hawaii. Her name was derived from cumulus cells,

cells that form a clustered layer that surrounds and supports the egg cell. Instead of transferring a nucleus into an enucleated ovum, Teruhiko Wakayama, a postdoctoral student, decided to insert an entire cumulus cell. The injected cell was then exposed to chemicals and growth factors to activate cell division. Since this first experiment, Wakayama has successfully created multiple generations of Cumulina mice — essentially making clones of clones.

The Charlotte generation of Project Leda followed the same process of making clones of clones. The first Leda generation was made from tissue and DNA samples from Kendall Malone (we'll learn more about her later), but these initial samples were lost in the lab fire, as Dr. Aldous Leekie explains to Cosima in season two. The original genome was lost. So, in order for the scientists at Dyad to create a second generation of clones, they needed to get new samples. And what better place to get such samples than from the clone already living at Dyad? Samples were taken from Rachel *four hundred* times in an attempt to make a viable second generation, but the culmination of all of this probing and sampling and experimenting was a single viable embryo: Charlotte Bowles. Turns out even Dyad scientists haven't perfected human cloning.

Dolly's and Cumulina's successes spurred scientists to clone a range of animals, from cows and pigs to horses and kittens. Some of these clones were created using SCNT, as Dolly had been, while others stuck with older processes, such as embryo splitting. By 1998, scientists at Kinki University in Japan were reporting success in using the Wakayama cloning technique with cows. Eight calves were cloned from one adult cow using cumulus cells. At that point, Kinki University's success rate was the highest in the world for cloning large mammals.

In 2005, Seoul National University was credited with cloning the first dog, dubbed "Snuppy," from a cell from the ear of a three-year-old Afghan hound. Snuppy the puppy was considered a particularly notable success because dog embryos had proved notoriously tricky to clone in the past. That said, Snuppy was the only survivor of 1,095 embryos. And Dolly was the first lamb to survive to adulthood out of 277 cloning attempts. In general, SCNT is inefficient — with the exception of cows, which for some reason have a slightly higher success rate when it comes to cloning (somewhere in the range of 5% to 20% efficiency, compared to 1% to 5% for all other animals). Scientists have managed to successfully create human embryonic stem cells using the "Dolly method" of nuclear transfer, but the success rate remains abysmally low thanks to poor blastocyst development. It makes you wonder just how many failures there were in order to produce the 274 human clones for Project Leda and the six (known) clones that we've met from Project Castor.

DOLLY BECAME (AND REMAINS TODAY) THE MASCOT FOR CLONING; SHE MADE HEADLINES WORLDWIDE AND EVEN GRACED THE COVER OF *TIME* MAGAZINE.

Dolly became (and remains today) the mascot for cloning; she made headlines worldwide and even graced the cover of *Time* magazine. But with that fame also came fear. Most countries hadn't considered any sort of legislation surrounding human cloning experiments. But if scientists could go ahead and clone a sheep without having to cut through all sorts of red tape, what was to stop them from doing the same thing with humans? By June 1997, just four months after

Dolly was publicly introduced to the world, the U.S. called for a federal law to ban human cloning and looked to develop regulations for existing cloning techniques. Interestingly, Ian Wilmut, the scientist who led the Dolly cloning experiment, has stated that human cloning "is an outcome I hope never comes to pass."

In January 1998, Richard Seed, a physicist in Chicago, announced that he planned to open a clinic to clone humans. He wasn't affiliated with any sort of reputable research facility and had no funding. If you dig into Seed's history you'll find that, outside of his Ph.D. in physics from Harvard, he had founded a company in the 1970s that specialized in transferring embryos in cattle, and later attempted and failed to found a venture using the same technique in humans. If this doesn't remind you in the slightest of *Orphan Black*'s Henrik Johanssen, the Prolethean religious extremist whose first episode shows him artificially inseminating a cow, then allow us to direct you to what Seed said in an NPR interview in 1997: "God intended for man to become one with God. Cloning is the first serious step in becoming one with God."

HENRIK

You see, I steered my faith through
science at MIT, and what I see here
is God opening a whole new door.

(2.02 "Governed by Sound Reason and True Religion")

Ultimately, Richard Seed had a few big media appearances before dropping out of the public eye. As far as we know, nothing has come of his claims. Since then, others have made claims of successfully cloning humans, but no one has stepped forward with a shred of evidence to prove these claims true.

Nowadays, a quick Google search will turn up more than a handful of organizations worldwide that offer human cloning services, some obviously fake and some apparently real ("real" in that they accept money for services, whether or not those services can be legitimately rendered). One of these services, Clonaid, has gotten more than its fair share of media attention, in part because of its close ties with Raëlism, a UFO religion that's been around since the

1970s. The foundation of the Raël movement is the belief that all life on Earth was created by extraterrestrials known as Elohim, and that these beings have been in contact with humans ever since our creation. Their interest in human cloning is essentially as a means to an end for eternal life. According to Clonaid's website (which, by the way, doesn't appear to have been updated since 2009), the first cloned human, dubbed Eve, was born in 2002 thanks to their services. Clonaid has never published any evidence of this success (or any of their apparently subsequent successes), citing the privacy needs of the persons involved.

More recently, attention has turned toward using cloning technologies to give second life to extinct animals. In 2013, a woolly mammoth was uncovered in Siberia, and it was in amazing condition — for being some 43,000 years dead. It was fresh enough for one of the scientists to take a bite out of its flesh (um, ew) and for the team to collect enough intact DNA to consider a cloning experiment. The mammoth genome has been sequenced from a few different mammoth samples, and recently scientists have taken to using powerful gene editing tools to insert specific mammoth genomic sequences into an elephant genome. At this cellular level, the animal that may be produced from these experiments wouldn't be a mammoth; it would be an elephant with some unusually mammoth-like traits. At the time of this writing, about 14 sequences have been introduced into the elephant genome in a lab setting. The next step would be to clone the elephant-mammoth cell, using the same process as was used for Dolly. Herein lies a challenge specific to elephants: the elephant hymen grows back between pregnancies to help the elephant carry its developing fetus to term (as you can imagine, elephants make for very large, heavy fetuses), and although it has a tiny perforation large enough for sperm to pass through, this perforation is not large enough for a developing elephant-mammoth clone embryo

A QUICK GOOGLE SEARCH WILL TURN UP MORE THAN A HANDFUL OF ORGANIZATIONS WORLDWIDE THAT OFFER HUMAN CLONING SERVICES.

to be inserted for implantation. Unless scientists figure out a way through the hymen, or they develop an artificial womb, we probably aren't going to be cloning mammoths any time soon.

These steps toward mammoth cloning aren't the first attempt at bringing an animal back from endangered or extinct status. In 2001, an Iowan cow gave birth to the first cloned endangered species: a baby gaur, or Indian bison. The baby gaur, called Noah, died quickly from a bacterial infection, but he demonstrated that one animal can successfully carry and give birth to a clone of an animal of a different species. In 2003, a team of scientists cloned a wild goat species known as the Pyrenean ibex, which had gone extinct in 2000. Scientists were able to use DNA from frozen tissues sampled from one of the last females of the species. Sadly, the cloned ibex died within minutes of birth because its lungs were malformed. While technically this experiment wasn't successful, it showed that de-extinction might be possible.

What does this have to do with *Orphan Black*? Well, the fact that we are even considering resurrecting a woolly mammoth shows just how far the technology has come since the 1950s. It shows how we can edit genomes and insert genes — just as was done to the Leda and Castor clones. It also shows us how many barriers would have existed when Dyad was undertaking their cloning projects and that many, if not most, of these barriers continue to exist today.

WHY BOTHER CLONING HUMANS?

The question of why the human clones should even exist in the first place is one that has not only been asked throughout the course of the show, but has been explored in science fiction since the 1930s, when Aldous Huxley's novel *Brave New World* was published. Human clones are often used in science fiction, like *Brave New World* or the 1967 novel *Logan's Run*, as a way to control populations and resources in a dystopian world. *Orphan Black* is set in a version of our society, and the show cites real and relevant events. Through that lens, and from a scientific point of view, there are some understandable goals that would lead a real group of scientists to pursue such a difficult and illegal plan of research.

In the season four episode "Human Raw Material," Susan Duncan — thought dead since the mid-1990s — is found to be very much alive and continuing human experimentation outside of Dyad and Project Leda with Neolution. She tells Cosima the truth about the Leda clones, a truth that even Ethan Duncan doesn't seem to know: that the clones were, in fact, designed to be blueprints for creating superior humans. Cosima is understandably pissed as hell about this revelation, and about the fact that it was Neolution toying with the human genome and inserting synthetic transgenes to create an improved human that caused the clone disease, which is killing her and her sisters.

COSIMA

Look at me, I'm sick. I never gave permission for any of this.

(4.05 "Human Raw Material")

At this point, Susan doesn't elaborate to say what the intended improvements to their genome might be, but any attempt to improve the biology of a living organism isn't exactly cut and dry.

So, the Leda clones are not only clones, but clones with synthetic sequences in their genomes, manipulations to improve on the original DNA. But is there any evidence that they are the next step in human evolution, stepping stones on the way to "perfect humans"? The scientists could have given the clones synthetic sequences that boost a human's natural abilities, such as rapid healing (Helena sure did recover from that bullet wound quickly), increased intelligence (Cosima, of course, is a genius, but each of the clones have exhibited quick thinking at some time), or super strength (we wouldn't want to get into a fight with any of the clones, but especially not with Alison). These synthetic sequences could be Dyad's first steps toward creating a human beyond the abilities of natural evolution, a human of heightened abilities that can bring on a new age of humanity. This avenue of cloning to create genetically altered transcendent human clones has been explored in the television series *Dark Angel*, as well as in *The X-Files* episode "Eve."

And then there's Project Castor, which approaches human cloning strictly for military purposes. It appears that the Castor clones are likewise intended to be superior humans — super soldiers that can be deployed like weapons — although what specific special or augmented abilities they might possess are not revealed. Super soldiers don't necessarily need to possess superhero-like qualities; they might merely be able to persist past normal human limits — a critical asset in battle. Perhaps the Castor clones can function for longer periods without sleep, or require less sleep than the average human. Maybe they have accelerated healing abilities, like we've seen hinted at with the Leda clones. Or maybe they don't have any special abilities, but the military wanted to create clones as expendable soldiers exempt from standard human rights protection (to keep "natural" humans out of dangerous missions or off the front lines of war). Any of these ideas pique your interest? You're not the first: check out works like *The Boys from Brazil* by Ira Levin, the *Star Wars* prequels and *Star Wars: Clone Wars*, and episodes from pretty much any sci-fi TV series from *Star Trek* and *Stargate SG-1* to *The X-Files* and *Doctor Who*. Expendable, reproducible humans are popular in science fiction.

But these are far from the only motivations for creating human clones. The scientists at Dyad could have created human clones simply to say that they could — and that would be plenty for most people. Scientific research is often driven by the desire to discover new things and achieve new goals for the sake of knowledge, and what better way to claim an understanding of the human genome than to successfully clone it in a set of viable human genetic identicals? In season five, we learn that Project Leda did continue as a study of the clones' behaviors and adaptations within different environments, but that seems like more of an offshoot of the original project to create living human clones than its primary purpose. A morally shady scientific corporation wouldn't need much more of a reason than being at the forefront of discovery to develop Project Leda.

Aside from pure bragging rights, studying the clones in different environments allowed Dyad to apply what they learned from the clones to advances in human health. By cloning the human genome and fostering a set of experimental subjects from birth, the scientists behind the project would gain many insights into the inner workings of the human body, from the relay of genetic code into observable traits (called phenotypes), to the viability of manipulated DNA as a potential therapeutic tool, to the effects of environmental factors on genetics. It's a scientifically beautiful setup to learn myriad things about ourselves as humans, and the doctors at Dyad were quick to jump at that opportunity.

LEEKIE

Understanding the human genome and its susceptibility to disease is the holy grail of life extension.

(5.07 "Gag or Throttle")

An extension of this human health potential is the idea of using cloned humans as living organ donors. It's practical: organ transplants from donors are risky because our immune systems are meant to destroy tissues that are foreign to our bodies. Right now, people who receive organ transplants basically need to weaken their immune system with drugs to increase the chance that their body will not reject the donated organ. The flip side is that weakening the immune system also means that organ recipients' immune defenses are down, and they are less able to fight off bacteria and viruses that cause illness. Perfectly matched organs from a clone would mean that fewer immunosuppressive drugs would be needed, and that the odds of the healthy organ being rejected would be reduced to practically nil. Breeding clones for their organs is popular in science fiction: Kazuo Ishiguro's novel *Never Let Me Go* (and its film adaptation) and Nancy Farmer's novel *The House of the Scorpion*, as well as the film *The Island*, have all explored the use of clones as perfect matching donors to (usually wealthy) people, in case they need an organ transplant.

ON THE ORIGIN OF *ORPHAN BLACK'S* SEASON ONE EPISODE TITLES

In a TV show where nothing is a coincidence and everything means something, it's no surprise that episode titles find their origins in thematic materials. For its first season, all the episode titles are phrases taken from Charles Darwin's seminal work *On the Origin of Species* (which we catch Cosima reading when she meets Sarah in a bar in season one).

Darwin's book changed the course of science with the creation of his theory of natural selection and his ideas on evolution. Natural selection refers to the differential survival and ability to reproduce among individuals of a given species due to phenotype — that is, the presence or absence of specific traits playing a role in an individual's ability to survive and pass on their genes. Darwin developed this theory while studying the finches of the Galapagos Islands: the various species of finches had each developed beaks specific to their diets and modes of feeding, and in *On the Origin of Species* Darwin discussed how this could only have been possible in these small island populations if a single original species had arrived at the islands and then developed these different phenotypes over generations of the finches adapting to different modes of feeding. In addition to all of season one's episode titles, *Orphan Black* gives a later nod to Darwin in season three's finale, "History Yet to Be Written," with a display of Galapagos finches in Rachel's Neolution holding cell.

Evolution and natural selection are important throughout *Orphan Black*, particularly in season one. While the viewers, along with Sarah Manning, are learning about the different clones and the scientists behind their creation, we see how these genetic identicals still maintain differences between each other, as well as how the traits they all share are adapted to each of their specific lives. For example, Cosima uses her intelligence in the world of academia, while Sarah uses hers to pull off cons, and Helena uses hers to aid in her assassinations. The clones all started from the same genetics, but they have adapted their traits to their environments in order to survive.

As for evolution, the entirety of Project Leda could be considered the next step in human evolution. Scientists manipulated each clone's genome in order to have a specific outcome, forcing the original genome to evolve in a specific direction. Dyad has taken on the role of the external forces deciding the fate of the clones' genomes. However, Kira represents an anomaly, an unexpected and unplanned outcome of their experiments, allowing the clone genome to have the ability to propagate beyond the original subjects.

To put it mildly, Darwin's work had a significant impact on the scientific world, an impact that paved the way for such disciplines as genetics and evo-devo and for such innovations as cloning. It is only natural, then, that this impact reverberates throughout the world of *Orphan Black*.

GENETIC IDENTICALS: CLONES AND TWINS

HELENA

We have a connection.

(1.04 "Effects of External Conditions")

Plot twist: Sarah and Helena are both Leda clones, but they are *also* twins. You've probably heard twins described as either identical or fraternal. Really, the appropriate terms are monozygotic (originating from one fertilized egg) and dizygotic (originating from two separate fertilized eggs). Sarah and Helena are monozygotic twins.

Does this mean that Sarah and Helena are more genetically identical to each other than to the other Leda clones? Well, yes.

How is it possible to be *more* identical? If they were the result of a

simple cloning experiment with no augmentation — that is, making a clone without manipulating or changing the original genome in any way — then there probably *wouldn't* be a difference between clones that are twins and the other clones. However, the clones' DNA also contains synthetic sequences inserted by scientists (which we'll discuss more in chapter four), and these sequences are in part what make Helena and Sarah more genetically similar to each other. As twins, they have the same ID tag sequence. The unique sequence was given to a single egg that was implanted into Amelia, Sarah and Helena's surrogate mother, and when this egg split in two, the ID tag sequence was then present in both clones.

There is also the infertility synthetic sequence to consider. Sarah and Helena can both bear children, while the other clones cannot, which means that part of their DNA coding for fertility must be different from the other clones'. This could be because of a mutation that occurred before the egg split in two, a lack of the sequence altogether in that egg, or some change in the regulatory elements (the parts of the genome that control the activation or repression of genes, but do not encode genes themselves).

HELENA

Scientists made one little baby
and then we split in two. So I
cannot kill you, sister. Like you
could not kill me. Sarah. We make
a family, yes?

(1.10 "Endless Forms Most Beautiful")

Of course, there are also random factors that could make Sarah and Helena more genetically similar to each other. If there were a random mutation in the genome, and if this mutation happened before they split into two embryos, then they would share this mutation, which the other clones wouldn't. Epigenetic factors (compounds that interact with DNA without changing the sequence, but can alter gene expression — we will go into this topic further in

---

CLONE CLUB Q&A

Do the clones have the same blood type?

Yes. Blood type is based on two things: the ABO type and the Rh factor (positive or negative), which are determined by your genetics. The clones would all have the same blood type. Your ABO group and Rh factor are both based on what antigens are found on the surface of your red blood cells.

Antigens play a huge role in your body's immune system: they trigger the production of antibodies in the blood plasma. Antibodies are proteins that will seek out and destroy harmful antigens (such as those from illness-causing bacteria and viruses, or from a blood donor whose blood type doesn't match yours). Individuals with type O-negative blood are known as the universal blood donors because the red blood cells have no antigens and so can be donated to any blood type without triggering an immune response.

chapter two) may also be shared between Helena and Sarah because they were carried in the same womb while the other clones were not.

Finally, there can also be *differences* between Helena and Sarah. A mutation could have occurred in one of the genomes after they split into two, giving one of them the mutation and not the other. There can also be epigenetic differences, as well as environmental differences, for each clone. For example, identical twins have different fingerprints because they experience the pressure in the womb differently, and it's that pressure that causes the ridges and folds that form your fingerprints. At any one of Helena's crime scenes, the blood sample and DNA fingerprinting analysis the investigators perform would result in a match between Helena and Sarah, but the actual fingerprints would not be *perfectly* identical.

Monozygotic twins also come with their own categorization system, based on how early the fertilized egg splits into two and whether the twins share a placenta, an amniotic sac, or both.

Once an egg cell is fertilized, it doesn't waste any time dividing over and over again; after about five days, it has gone from being a single cell to becoming a ball of 200–300 cells, known as a blastocyst. Most monozygotic twins result from a split at this stage (between four and eight days after fertilization). These twins usually share a placenta but have separate amniotic sacs: they are sharing their nutrient, blood, and air supply (not to mention waste elimination), but they are each growing in their own protective environment. If the split happens before the blastocyst stage, the twins will likely each grow with their own personal placenta and amniotic sac.

Helena and Sarah are most likely the rarer type that share the same placenta and amniotic sac. About nine days post-fertilization, the blastocyst has implanted in the uterine wall and has begun to define body planes. Body planes are lines that we can hypothetically draw to section a body. For example, to imagine a midsagittal plane, picture a line being drawn from the top of your head down the middle of your forehead, between your eyes, and all the way down the midline of your body, such that it divides the right side of your body from the left. A coronal plane would divide the front of the body from the back (so imagine that dividing line starting at the top and running down the sides of your head, through your ears), and a transverse plane creates horizontal sections, like slicing a loaf of bread. Anatomists use these planes to map out where different parts of the body can be located. So, when the blastocyst splits along a plane to form what's known as mirror twins, the twin on the right would need to generate a new left side, and vice versa. If a blastocyst splits any later, there is a risk of conjoined twins.

Sarah and Helena are monozygotic mirror image twins, meaning that they developed from a single fertilized egg that split into two embryos during the early stages of development. This explains, for example, why Helena is left-handed while Sarah is not (and Helena is the only known left-handed Leda clone). The split resulted in the embryos having mirror image body plans. Note that body *plans* are different from body *planes*. A body plan is the mapped out physical placement of all the body's parts and organs, whereas planes are lines of division. If we consider their body plans, Sarah and Helena don't

CLONE CLUB Q&A
If Sarah and Helena are supposed to be "mirrors" of each other, why wouldn't the little mole on Sarah's right cheek be on Helena's left?

Mirror twins do usually have birthmarks opposite one another, just as they have opposite handedness and opposite hair whorls (and, in the rare case of Helena and Sarah, inverted organs). There are cases, however, of only select traits being mirrored.

Of course, having the mole switched would also make it much easier for us, the audience, to figure out clone swaps (like when Helena-as-Sarah encountered Paul at the end of the season one finale, "Endless Forms Most Beautiful"), and that would just spoil the fun.

But, yes, as far as the biology goes, if they are perfect mirror twins, they should have moles in the opposite position, not the same position.

just have opposite handedness: Sarah ended up with her organs in the normal positions, while Helena's organs are flipped. This is typical of a condition known as situs inversus. Literally meaning "inverted position," situs inversus is a condition in which the locations of the internal organs of the body are reversed, creating a mirror image of their normal locations. Sometimes a single organ is affected: a person with dextrocardia has their heart located toward the right side of the chest instead of the left. For people with situs inversus totalis, all their organs are flipped: their hearts are located on the right side of their chest instead of their left, their livers on the left side instead of the right, and so on. This condition can occur as the result of a recessive genetic condition passed on to the child from two carrier parents, or it can occur in mirror twins.

Situs inversus is a very rare condition, occurring in only one of every ten thousand births. During normal heart development, the cardiac tube arises from an outgrowth of the middle layer of the embryo that folds and rotates, eventually giving rise to the four chambers of the heart. In the case of situs inversus, the tube rotates improperly so the right chambers end up on the left side of the body, and vice versa. This mislocation is then propagated throughout the rest of the body, creating the mirror image body plan.

Most often, situs inversus goes undiagnosed until an injury or illness occurs, because there are usually no side effects from having the organs on the opposite side of the body. In Helena's case, her situs inversus is not discovered until Sarah shoots her point-blank in the chest. Such a wound would be fatal for most people and would have certainly killed Helena had she not been Sarah's mirror twin. The bullet did not hit her heart because her heart was not in the expected location, and therefore Helena could limp over to the hospital in one piece . . . with the help of some duct tape and a belt.

TOMAS

```
Though one could easily mistake
the bullet missing her heart for a
miracle.
```

HENRIK

It is a miracle, my friend.
Your charge is a genetic anomaly.
She's a mirror.

(2.02 "Governed by Sound Reason and True Religion")

While situs inversus ended up saving Helena's life, some clones' developmental side effects have proven to be detrimental, even fatal. Such is the case with the second generation of Leda clones.

THE CHARLOTTE GENERATION

Although Ethan and Susan Duncan's lab at Dyad was destroyed in an explosion in 1990, supposedly along with all the notes and materials crucial to creating clones for Project Leda, Dyad did not let this loss end their project. In the season two finale, we meet Charlotte, a Leda clone from the same cell line as Sarah Manning but with one significant difference: she is only eight years old. According to Marion Bowles, a high-level Topside executive and Charlotte's adoptive mother, Charlotte is the sole survivor of 400 attempts to create a clone in the 20 years since the fire that destroyed the Duncans' lab. Charlotte is the second generation of Project Leda.

As mentioned earlier, during the process of somatic cell nuclear transfer (SCNT) cloning, the nucleus of a cell from the organism to be cloned is removed and placed into a donor egg cell. This egg is then activated to behave like a newly fertilized egg cell — now called a zygote — and it gives rise to a new, cloned organism. However, there is a key problem in this procedure — what Ethan Duncan refers to as the "spindle protein problem" — that leads to complications. It's the most likely culprit for the problems that made the Charlotte generation so difficult.

There are many steps that lead a single cell to divide into two identical daughter cells. During one step known as metaphase, the nuclear membrane disintegrates and the duplicated chromosomes line up at the center of the cell, a location known as the metaphase plate. At the two ends of the cell are structures known as the mitotic

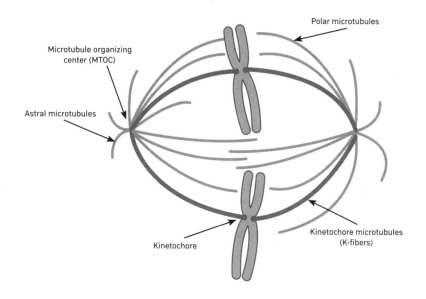

Polar microtubules

Microtubule organizing center (MTOC)

Astral microtubules

Kinetochore

Kinetochore microtubules (K-fibers)

The spindle apparatus regulates the outgrowth of microtubules to the chromosomes, which then attach and help divide the chromosome pairs into two complete sets of DNA on either end of the cell. When this process is disrupted, the result is irregular division of the chromosome pairs, which can lead to cell death.

spindles, which are made up of microtubules (long tubes of protein that help maintain a cell's shape) and other structural proteins. The microtubules reach from the spindle to the metaphase plate and connect to the centers of the chromosomes. This cues the next step, anaphase, when the microtubules pull one chromosome of each duplicated pair toward the spindle, separating the two halves and creating two clusters of chromosomes that eventually provide the genetic information for the two daughter cells.

During the process of SCNT, activation of the egg cell can occur at the wrong time if the spindle apparatus is stimulated. Improper cell division then leads to chromosomal abnormalities that are potentially fatal. This problem plagued scientists developing cloning techniques, and it's probably what caused 399 of Dyad's 400 attempts to make new clones to end in fetal termination.

In 2013, a team of scientists developed a modification of the SCNT procedure that involved removing the spindle apparatus from

the egg cell and replacing it only after the donor nucleus had been transferred into the cell. This technique prevented spontaneous cell activation and in effect solved the "spindle protein problem." Most likely, in the world of *Orphan Black*, Susan Duncan came up with a similar solution back in the early 1980s, allowing the Duncans and their team to successfully create human clones. Because she hid her notes from Dyad, the second generation of Leda clones was such a struggle to create.

> **ETHAN**
> The military recruited Susan and
> I in '76. Susan was the brains,
> really. Did you know that she
> cracked the spindle protein problem?
>
> **SARAH**
> What did you do?
>
> **ETHAN**
> We succeeded. Cloned human embryos.
>
> *(2.06 "To Hound Nature in Her Wanderings")*

Charlotte managed to survive embryonic development despite the presence of the egg's spindle apparatus for the entirety of the process. During development, her cells were most likely activated

CLONE CLUB Q&A.

Why would Katja's prints flag a match to Sarah's in season one? Aren't fingerprints created in the womb and not based on genetics?

This is a scientific inaccuracy in *Orphan Black*, but it was a necessary one from a plot standpoint: Art needed to find out about Sarah at the end of "Parts Developed in an Unusual Manner" once Sarah-as-Beth quit the police force — even though Sarah thought she'd covered her tracks when Katja's prints first flagged a match to her criminal record in "Variation Under Nature."

The fingerprints of identical twins have many similarities — much more so than any other two people — but they are still distinct. Fingerprints are formed in utero by stresses placed on the dermal cells (the layer of skin below the epidermis but above the subcutaneous tissue). These stresses are different for each twin; they're environmental, not genetic. All the clones, save Sarah and Helena, developed in different uterine environments — those of their surrogate mothers.

All the clones should have distinct fingerprints, despite being genetic identicals. Environmental wear and tear also causes changes in fingerprints, so even if the clones all had the same fingerprints at birth, they should be distinct by the time they're in their late 20s.

It's possible that some elements to fingerprints may be heritable, and most fingerprint databases do select potential suspects that are partial matches to the fingerprint evidence. So, while there is a chance that Sarah and Katja might have enough similarities to flag a database match, it would be a partial match, and not the exact match that we saw in the show.

improperly by the egg's spindle apparatus, leading to improper cell division and eventual chromosomal disruptions and abnormalities. These problems, however, happened to not be fatal in her case, and instead just led to developmental differences and physical limitations, such as the necessity for a leg brace.

While Sarah and the other clones her age are treated like they are just one of a vast number of replicas across the world, Charlotte is treated like a miracle, yet they all are genetic identicals of each other. Roughly two decades span the time between when the original Project Leda clones were created and when Charlotte was born, yet the circumstances couldn't have been more different.

When Sarah Manning stepped out onto the train platform at Huxley Station and first laid eyes on Beth Childs, she was seeing so much more than someone who looked just like her. She was seeing the culmination of decades of progress in science and biotechnology and of countless obstacles surmounted to create a cloned human being. And that's just the beginning. This chapter has mostly focused on biological hurdles; it has barely scraped the bioethical implications of creating human clones. For that, we'll need to start a new chapter . . .

CASE STUDY:
SARAH MANNING

ID: Unknown
DOB: March 15, 1984
Birthplace: London, England,
United Kingdom
Status: Alive

COSIMA

You're the wild type, Sarah. You
propagate against all odds. You're
restless. You survive.

(2.10 "By Means Which Have Never Yet Been Tried")

When Cosima holds Sarah's hand and calls her "the wild type," she's not just referring to Sarah's uncombed mane of hair and penchant for punk rock. In genetics, the term "wild type" is used to refer to a non-mutated, or "normal," allele of a gene. (An allele is one of different forms of a gene.) In terms of phenotype, or expressed genetic trait, the wild type trait is the trait most commonly found in nature.

This is sort of a simplified explanation of genetics from when it was still developing as a science. Gregor Mendel, a mid-19th-century Austrian monk known today as the "Father of Modern Genetics," did not coin the term "wild type" — and in fact didn't even know what genes were — but he did develop important experiments that built our understanding of trait inheritance. His experiments involved breeding pea plants to see what traits got passed down through generations. He chose to track seven traits that were easy to identify and that apparently had only two forms in the plants that he was studying. As an example, one of the traits that he studied was pea color. In peas, there is a gene for seed color, and it has two alleles that give two different phenotypes: yellow seeds and green seeds. When Mendel bred one pea plant with green peas and one pea plant with yellow peas, the resulting generation of plants always had yellow peas. If he bred the yellow pea daughter plants together, the green peas would reappear in the next generation, always with a ratio of three yellow pea plants for every one green pea plant. Wild type is often used to reference the most prevalent allele as it would be found in nature. In the case of peas, the wild type allele is for yellow peas; the other allele of the gene (green) would be referred to as the mutant allele. Mendel grew over ten thousand pea plants over the course of eight years. He kept meticulous notes about the traits of each one and found that the same pattern emerged for trait inheritance for all seven of his studied traits. Today, when studying mutations and patterns of trait inheritance, scientists turn to bacteria or fruit flies because they are easy and fast to breed, so multiple generations can be observed more efficiently. A wild type fruit fly, for example, has red eyes and a brown body (all wild type fruit flies look more or less the same). A mutated fruit fly might have white eyes, amber eyes, or no eyes at all; its body might be pale, or it might produce too much pigment; it might have curly wings or shortened wing nubs. Next time you spy fruit flies in your kitchen, take a closer look and see how many have wild type traits and how many are mutants.

Cosima tells Sarah that she is the wild type because she possesses traits that the other clones do not: she is fertile, while the other clones have been mutated to be infertile. Sarah propagates against all odds. She can reproduce despite all of the efforts by Duncan and his team to make the clones barren by design; she is restless, she survives. The

purpose of reproduction in nature, in general, is to have your genetic information survive to be passed on to future generations, and Sarah has achieved that by having Kira. Sarah also does not possess the mutation that causes the clone disease because it was introduced when the infertility sequence was inserted into the Leda genome. When synthetic sequences are spliced into DNA, it is difficult to precisely target exactly where they will insert themselves, and they can disrupt or alter the expression of other genes. This is true even today and would absolutely have been true when Project Leda was modifying DNA in the late 1970s and early 1980s. Unlike the other Leda clones (excepting Helena who, as you might've guessed, is also the wild type phenotype), Sarah is immune. As she tells Mrs. S in "From Dancing Mice to Psychopaths," "It's my curse to watch my sisters die."

And so it's a bittersweet moment when Cosima tells Sarah that she won the genetic lottery — she somehow escaped mutation, she reproduced, and her genes survive, while Cosima and most of the other clones are unable to propagate and are dying.

SARAH

Do you think, because Kira's mine,
she might be different?

COSIMA

Because she's uninjured?

SARAH

Yeah, it's just a question.

COSIMA

Well, if they genetically modified
us, I mean, you could've passed
something down to her.

(1.09 *"Unconscious Selection"*)

Food for thought: typically, in lab experiments studying gene mutations, you run your experiments with a control parallel to the variables. Oftentimes, this control is a wild type version of your gene of interest, so you can compare the mutants back to the unmutated form. Despite Ethan Duncan's insistence that the clones were all intended to be infertile, and that Sarah's and Helena's fertility is a sort of genetic fluke that allowed for them to revert to wild type, it's entirely possible that the research teams set aside some wild type clones to study. Project Leda would have been a landmark experiment, and the scientists involved would have been remiss if they didn't take every opportunity to learn about the genetics of clones. If twin studies have taught us so much about heredity and factors of nature versus nurture, then clone studies would surely serve at the very least to enhance this learning. And the fact that Sarah is fertile and has given birth to Kira means that some of Sarah's traits can be observed in a daughter generation. Given Cal's genetic contribution, an unknown in terms of Kira's genetic makeup, it isn't exactly a controlled experiment — but perhaps Sarah's wild type status isn't as much of a genetic accident as everyone thinks it is.

CASE STUDY:
ALISON HENDRIX

ID: 903V18
DOB: April 4, 1984
Birthplace: Scarborough, Ontario, Canada
Status: Alive

ALISON

I tried to say "eff it" today and I blew up my whole life. I just wanted to say "eff this," "eff you," and I effed it, I effed it all up.

(*1.08 "Entangled Bank"*)

Alison Hendrix is, more so than the other clones, defined by how she wants others to see and perceive her. From the moment that we meet her and throughout the seasons, she is constantly working to project and protect her ideal of what a perfect mother of a perfect family should be: soccer and figure skating coach, active in community theater as a seamstress *and* actor, Glendale District School Board Trustee, and giver of handmade gifts and baked goods. In these respects, Alison Hendrix is a model citizen who must make solid choices to shoulder this much responsibility. But for a suburban soccer mom concerned first and foremost with protecting her home and family from everything unfolding around her, Alison Hendrix sure can't seem to resist making some awfully risky choices. Alison is the quickest to draw a gun to deal with a situation. We've also seen her knock her husband out with a golf club, tie him up in her craft room, and threaten him with a glue gun practically the moment she suspects he is her monitor. Anyone else might have waited things out until they could build enough evidence to be convinced one way or the other and evaluate the safest way to deal with a monitor. Not to mention that Alison's big solution to her money problems is to start a neighborhood drug "entrepreneurship" through her mother's soap business. Finally, we've seen, through her various dependencies on alcohol and prescription drugs, that Alison is prone to addiction, and the rush that she gets from risk-taking can be just as addictive.

It all ties back to the most important reward system in the brain: a nerve pathway known as the mesolimbic dopamine system, including a deep structure in the brain called the nucleus accumbens. This part of the brain is responsible for signaling for a rush of dopamine, which is sort of a celebrity among neurotransmitters for its association with sex, drugs, and addiction. Of course, it's not as simple as news articles make it seem when their headlines announce that Oreo cookies are as addictive as cocaine because they trigger the same surge of dopamine in the brain. Dopamine is a neurotransmitter that can make you feel *good*. Or, at least, it plays a role in feeling good. But that's not all dopamine does, even if that's all it gets press for. It acts on several different receptors to trigger different responses and has more than one form. Dopamine absolutely plays a role in addiction, but it also plays a role in initiating movement. Destruction of dopamine-producing cells

in a part of the brain known as the substantia nigra is responsible for the disordered movement seen in Parkinson's disease. Dopamine also has a hormonal function and stops the release of breast milk (whose release is triggered by a separate hormone, prolactin), which is why some anti-psychotic medication that alters dopamine levels may trigger milk production, even in men. It's also involved in attention and salience (noticing that something is important and requires attention). While the mesolimbic system sees spikes of dopamine release with reward, it also sees increases in dopamine levels in people with post-traumatic stress disorder when they are experiencing stress and paranoia. Dopamine and the mesolimbic system are also implicated in psychosis.

In terms of risk-taking and addiction, the nucleus accumbens in the mesolimbic system is the part of your brain that tells you to pay attention to exactly what it was that gave you that feeling of pleasure, and it reminds you of the rush you felt when you won on a gamble or ate a cupcake, motivating you to seek out the next pleasurable rush. Evolutionarily, the nucleus accumbens and the reward system are important because they guide and reinforce behavior essential to survival. (For example, eating good food is pleasurable for most people and thus motivates us to seek out more good food.) The tricky thing is that appetite for risk-taking is variable from person to person.

ALISON

Oh, I'm fine. It's my decision. Everything is under control.

SARAH

Okay, now I'm worried.

(*1.08 "Entangled Bank"*)

Most adult humans are pretty risk-averse. If presented with a risky option with a high payoff versus a low-risk option with a lower, but guaranteed, payoff, they will usually go for the latter. Alison? Well, she may not ever admit to it, but she seems to prefer risk. Most addictive drugs are addictive in part because they trigger a release of dopamine. If a risky choice pays off in a pleasurable way, it will likewise trigger a release of dopamine and become addictive. The brain will remember this rapid sense of satisfaction and create a conditioned response to whatever provided that pleasure. Some studies have even found that dopaminergic neurons will fire when pleasure is *expected*, whether that pleasure actually happens or not, which suggests that dopamine might have a role in desire for pleasure as well, linking to its role in motivating behaviors. The jury's still out on what factors — whether genetics, personality, or stress — lead to predisposition for addiction and risk-taking behaviors.

Alison isn't the only clone prone to addiction: we see this tendency in Beth as well and learn about Sarah's past as an addict in seasons four and five. And Alison *certainly* isn't the only clone who has made risky choices: Cosima nearly implanted a maggot-bot in her cheek as a Hail Mary cure for her disease, and Sarah's first impulse upon seeing Beth walk onto the tracks was to *steal her purse from the platform*. That said, for the most part, we can see the stakes that pushed these impulses through. But Alison? Holy doodle, she sure has a knack for taking relatively stable situations and destabilizing them with unnecessarily risky behaviors.

"THERE'S ONLY ONE OF ME"

NATURE VERSUS NURTURE

FELIX

Soccer mom Sarah? Dreadlocked
science-geek Sarah? Arguably more
attractive than the real Sarah.

(1.03 "Variation Under Nature")

So, let's say you, like Sarah Manning, find out that you're part of a cloning experiment. You're one of who knows how many human clones. They look exactly like you; they talk exactly like you. Maybe they're even wearing the same clothes as you are. After all, their brains are forged from the exact same genetic material as yours — it only follows that they'd have the same taste in T-shirts. You're all clones of the same cell line, so you must all be exactly the same.

Right?

It may sound ridiculous, but for some reason there's this expectation with clones that, because their cells contain perfect copies of the same DNA, they must then be perfect copies of each other. It's become a trope: clones are carbon copies, indistinguishable from each other in form and thought, practically able to finish each other's sentences. Clones of famous figures, good or evil, surely must grow into the very traits for which they are famous. But real-life clones — monozygotic twins — aren't expected to be exactly the same. Each is expected to develop their own distinct interests and mannerisms despite being nearly genetically identical and despite (in most cases) being raised in very similar environments. Why should we imagine anything different for clones?

SARAH

So is this the part where 20 more
of you robot bitches walk in for
effect?

(1.10 "Endless Forms Most Beautiful")

The entire premise of *Orphan Black* stands on clones who are individuals, as human clones would realistically be. You can instantly tell Cosima from Alison or Sarah, not only by their hairstyles and accents, but also, and arguably more importantly, by what motivates them, by their stances on issues, by the choices that they make, by how they react to and interact with the world, and all their other little idiosyncrasies. If you put Rachel in a wig, you don't get Sarah. (Although Rachel puts on a performance strong enough to fool at least Mrs. S for long enough to kidnap Kira. Maybe not quite as good as Sarah's chameleonic ability to imitate her sisters, but miles better than Cosima's single, fumbling attempt at delivering a school board trustee election speech as Alison. Inherited traits might lend themselves toward building certain skills, but talent is individual.)

So, how does personality form? First, let's define what we mean by personality. Looking to the word itself doesn't reveal much: *personality* is derived from the Latin *persona*, meaning a mask or portrayed character. Is personality then the "masks" that we wear in any given situation? This seems to imply that personality is a choice. Behaviorists like Albert Bandura and B.F. Skinner have studied personality as a pattern of behaviors learned through observing and repeating others' behaviors. They also suggest that those behaviors that form personality are reinforced through reward and punishment. Is personality then a collection of traits that we've learned

from others applied to a blank slate? Can personalities be trained? Other studies have correlated personality traits to neurotransmitter activity in specific areas of the brain; neuroticism, for example, has been linked to activity in the anterior cingulate cortex, an area of the brain also associated with human bonding, emotion, decision making, and reward anticipation. Is personality then nothing more than a cascade of brain chemicals and firing neurons? Clearly, personality is a complex combination of factors. It's a relatively consistent way an individual person relates to the world around them: emotions, motivations, learned behaviors, and intrinsic preferences. How do you even begin to measure that?

Personality can be defined as the aspects of human nature and human behavior that unite us all as people (the nomothetic perspective); it can also be defined as the elements that characterize us as individuals (the idiographic perspective). Around 400 BC, Hippocrates and others looked to a balance of humors within the body to inform personality and temperament. Humors were basically bodily fluids — black bile, yellow bile, phlegm, and blood — which were thought to influence temperament and health. Some personality traits were linked to an imbalance within these four humors that strongly favored one over the others. An ambitious, hot-tempered, leader-type person might have been described as choleric, with an excess of yellow bile, while a quiet, serious type of person might have been described as melancholic, with an excess of black bile. It was . . . far from a perfect system for describing personality. In the 19th century, German physician Franz Joseph Gall's popular practice of phrenology looked to skull size and shape to provide insight about personality: any bump or valley in the skull bone might be associated with a different trait. While today neither phrenology nor humorism is given much credence as far as personality development is concerned, both have been considered historical stepping stones to contemporary study of the subject.

In 1884, Sir Francis Galton (who coined the phrase "nature versus nurture," not to mention the term "eugenics") was one of the first to set out to create a taxonomy of human personality traits. He did this by looking at language and categorizing words according to how they described personality; his theory was that individual personality traits must be encoded in the language that we use to

CLONE CLUB Q&A
Dyad didn't expect Sarah and Helena to be twins, did they? Do they share the same tag number then, or is it possible to tell them apart?

Sarah and Helena have the same tag number in their DNA. Kira also has this same ID tag sequence.

Dyad doesn't have any samples from Helena because she was always out of their hands, and they have limited samples from Sarah, obtained the few times Dyad has managed to pin her down. They haven't yet had a need, or even enough material, to distinguish their samples.

Of course, Henrik did give a sort of tag number to Helena — E302A — which can be seen on blood vials from the Prolethean Ranch in "Things Which Have Never Yet Been Done," but it doesn't align with the Project Leda naming system (e.g., Cosima's 324b21).

describe people. He chose about 1,000 words for his dictionary of characteristics; others who carried on with his work decided anywhere between 750 and 3,000 words were important in defining personality. Imagine trying to do a personality quiz where your result is a list of 1,000 traits. Personalities are nuanced, but Galton's taxonomy is overkill.

At least when it comes to personality research, many studies look to what's known as the Big Five, a personality model that divides traits into five major categories: openness to experience (curious versus cautious), conscientiousness (efficient versus easy-going), extraversion (outgoing versus introverted), agreeableness (friendly versus antagonistic), and neuroticism (nervous versus confident). The Big Five model is generally used to measure personality by asking people to self-report where they fall within the different scales, such as by ranking the degree to which they agree or disagree with statements such as "I feel comfortable around people" or "I pay attention to details." It's not the only model, but many personality measures base themselves on the Big Five at least in part. The Minnesota Multiphasic Personality Inventory looks at many more traits and is the most widely used for psychological testing today; one of its sections basically looks at the Big Five traits, adjusted somewhat more in accordance with its clinical bent. If you've ever done a personality test online, like the Myers–Briggs Type Indicator (MBTI), the test's scales were likely also based on the Big Five model.

The Big Five model lets us assign personality traits to ourselves and others. Alison, for example, would score highly on neuroticism and conscientiousness, and maybe somewhat less highly on agreeableness. Cosima, easy-going as she is, would score much lower on conscientiousness and higher on agreeableness and openness to experience. But this model doesn't get to the crux of why these traits exist or where they came from, and why certain traits appear in some of the clones and not others.

Galton's work investigating personality involved a questionnaire, which he built and sent out to 190 scientists (specifically, Fellows of the Royal Society, which has collected some impressive members throughout history, not to mention good ol' "Father of Neolution" P.T. Westmorland). Galton looked at the fellows' birth order, parents' occupations, race, and other factors to determine if there was a correlation between these factors and the fellows' interest in science. He raised the debate about nature versus nurture, but he did not settle it. Since Galton's study, scientists have worked to determine to what extent traits are influenced by genetics and biology as opposed to external factors such as environment and other people.

COSIMA

Bad brain chemistry can be genetic, um, but environment, that's individual, right? I mean that's the whole nature/nurture question right there.

(1.04 "Effects of External Conditions")

These days, the predominant view is that nature and culture are both significant factors that form our traits, personalities, and behaviors. It's a pretty reductionist approach to think that only one factor, either genetics or environment, could be wholly responsible for something as complex as personality. As psychologist Donald Hebb once anecdotally remarked, "Which contributes more to the area of a rectangle: its length or its width?" Many traits — and, in fact, all the Big Five traits — have been linked to a hereditary component. But a given genotype, a trait encoded in a person's genes, might lead to a certain phenotype, or expressed trait, only under the right environmental circumstances.

The best way to study nature versus nurture, of course, would be to study people who are as genetically similar as possible to measure just how similar (or different) they end up. In the absence of human clones, the classic tools for this sort of research have been twin studies and adoption studies — twins raised together or raised apart. These studies look for discordance: traits that show up in one twin but not the other. Given their nearly identical genes ("nature"), any discordance becomes a point to examine more closely for evidence of the influence of social and physical environment ("nurture"), whether that environment is shared by both twins or not. Probably the most well-known collection of twin studies — the Minnesota Twin Family Study — followed monozygotic and dizygotic twins who were separated at an early age. The study ran from 1989 to 1999, studied 137 pairs of twins (81 of these pairs were monozygotic), and was made up of several smaller studies focusing on certain traits. One famous pair, Jim Lewis and Jim Springer, monozygotic twins separated at four weeks old and reunited at age 37, were found to be uncannily similar: they smoked the same brand of cigarettes, drove the same car, vacationed at the same Florida beach, and both had a nail-biting habit, among other surprising similarities. Most of the twin studies *did* find that there was a stronger incidence of the same traits across monozygotic twins than dizygotic twins (which is not surprising, really, when you consider that dizygotic twins share only about 50% of the same DNA, as would non-twin siblings). Many of the studies also found similar traits in both twins even when raised separately, suggesting a genetic, inherited component. One of these examined traits was religiosity: identical twins were found to be either both religious or both non-religious more often than unrelated individuals. If we think about the Leda clones, Helena and Alison have both been shown to be religious — could there be a genetic component? That said, Sarah is more genetically similar to Helena than Alison is, and Sarah has never shown any sort of religiosity; if anything, she seems dubious of religious practice and faith. If we consider Helena and Alison, we can note that they were both raised from a young age in religious environments, Helena in a strict Ukrainian convent and Alison in a Christian household. Is their religiosity strictly a product of these environments? Or did these environments simply foster the expression of existing, inherited traits? It could be argued in this case that Helena's religiosity is heavily environmental. When she's introduced to the series, she is described as

CLONE CLUB Q&A
I understand that accents and dialects are a social/environmental thing, but each clone's voice seems to have a different pitch/tone. Is that also social, or is there a biological explanation as well?

Biology and environment both have an impact on your voice pitch and tone. Your throat size, mouth size, and tongue all play a role in voice tone, and those are determined by genetics — so they'd be the same for all of the clones. But the way you breathe is a learned trait that is influenced by your environment. Alison has a higher-pitched voice because she has a tighter way of breathing than Sarah, who breathes from deep down in her diaphragm. If you try singing different high and low notes, you will feel how your breath seems to sit in different positions in your body.

The only Leda clone who would display real differences in the anatomical aspects of voice production is Tony. One of the physical changes that comes with taking testosterone injections is that the vocal folds (also known as the vocal cords) in the larynx become longer and thicker. These vocal folds vibrate air to produce the sound of your voice; since Tony's vocal folds will be longer and thicker than the other Leda clones', they will vibrate differently and produce a lower tone. Otherwise, the clones' voices are all similar enough, with any differences resulting from their environments and individual speaking habits.

a religious zealot, but she never mentions God or prays again after breaking with Tomas (although in season four, we do see her place a cross over her buried embryos, as well as decorate her hut with religious images). It's likely that most of Helena's religiosity was born as a survival behavior — conforming to the rules of the convent and to Tomas's rules to avoid abuse. Her beliefs were beaten into her. If Sarah had gone to the Church instead of Helena, would she have shown the same or similar beliefs?

HELENA

You gave me to them. You let them make me this way.

(1.10 "Endless Forms Most Beautiful")

And while twin studies have shown that traits tend to appear more often across twin pairs than non-twin siblings, they have also found in many cases that environment is a significant factor, especially when twins are raised apart, such as is the case for the Leda clones.

RACHEL

I'm not my sisters. I'm Neolution bred.

(4.07 "The Antisocialism of Sex")

Rachel Duncan, pro-clone and "uber-bitch," as she is dubbed by Sarah, is the only first-generation Project Leda clone known to have been raised self-aware. At only six years old, she was brought from Cambridge, U.K., to Dyad headquarters in Toronto, Canada (after the lab fire and her parents' disappearance), where she was set up with a new bedroom in the facility. On the bed was a bound copy of the research that led to her creation.

What would Rachel Duncan be like if she had been raised a naive clone? How would the other clones be different if they had been raised self-aware? Rachel is ambitious, driven, and aggressive. She is cold and selfish — Machiavellian, even. To what degree are these traits intrinsic to Rachel's personality, and to what degree are they born of her upbringing at the Dyad Institute, an organization that framed Rachel as a Chosen One, raised her up, and set her aside as Other, going so far as to give Rachel power to affect the other clones. Rachel's upbringing at Dyad also placed pressure on her to be not only a successful businessperson within the organization but a model clone specimen. We see important figures in Rachel's life, notably her mother Susan Duncan, vocally expect Rachel to be more than the other clones, downplay any successes, and criticize any perceived failings, all while maintaining Rachel's status as Other, separate from the surviving Leda clones. In season four, Susan repeatedly tells Rachel that she is a disappointment despite being raised self-aware and living a life guided heavily by Neolution. Such an environment would certainly have shaped Rachel's personality and behaviors.

Other clones, like Alison and Helena, have shown similar tendencies for approval-seeking. If either had been raised self-aware in Rachel's stead, would they make the same choices? In the same vein, how would Sarah and Helena be different if Sarah had gone to the Ukrainian convent and Helena had gone into the foster system? Would Sarah still be a rebellious force? Would Helena find a different cause to dedicate herself to with fervor? The nature versus nurture debate becomes an interesting thought experiment once you start to consider which of the clones' traits might be inherited and which might be a product of environment. Members of Clone Club have explored these questions through discussion and fiction, and as many theories as real possibilities have developed. It's difficult to predict to what degree "nurture prevails," as Rachel states when she is moved to spare Dr. Leekie in "Knowledge of Causes, and Secret Motion of Things." As far as science goes, current sentiment is that, while it's unclear if one dominates over the other, both nature and nurture are important factors informing personality and behavior.

RACHEL

When I was six years old, I became
the only Leda aware of the other
clones. Dr. Leekie shows me their
files so I can understand how
experiences determine traits.

(5.07 "Gag or Throttle")

One especially fascinating element to Rachel is that her Otherness has led her to make decisions that affect her as much as the other clones. In the season two finale, "By Means Which Have Never Yet Been Tried," she crushes Kira's stem cells underfoot to prevent Cosima's treatment, when it is in Rachel's best interest for Cosima to be healthy so that she can produce a cure for the clone disease. Rachel does this despite knowing that the clone disease is encoded in her DNA just as much as it is in Cosima's, and that finding a cure for Cosima also means finding a cure for a disease that she will likely develop. This behavior could suggest either that Rachel is so highly motivated to hurt Sarah that it trumps finding a cure, or that she is confident that the other scientists at Dyad will develop a cure without Cosima — and she would rather have a solution that doesn't indebt her to a clone. But then in the season four finale, "From Dancing Mice to Psychopaths," Rachel introduces a new conflict: she proposes launching new human cloning experiments in countries where human cloning is not illegal and where Neolution's "corporation supersedes [the clones'] citizenship, their personhood." What Rachel fails to realize, or more likely ignores, is that in these countries she is likewise stripped of personhood. It's more than an understatement to say that Rachel shows an unusual lack of affinity for her clones. One of the Minnesota Twin studies surveyed twins raised separately to get a sense of how they felt about their twin once they were finally reunited. Among the monozygotic twins surveyed, about 80% reported feeling close and familiar with their twin, a bond even closer than that which they had with their closest friends. This particular study suggested a genetic component to the bond felt between monozygotic twins, genetic identicals. Apparently, Rachel Duncan missed that memo.

The Castor clones, in contrast, were raised in an environment that heavily favored similarity. Apart from Ira (who came into Susan Duncan's care and only spent his first years of life with his brothers), the Castor clones were raised in a structured military environment in close contact with each other and aware that they were genetic identicals, which absolutely contributed to the tight-knit nature of the Castors. In terms of familial bonding, they are unusually close. And although they may be somewhat more challenging to distinguish from each other, they do have distinct and separate personalities.

THE CASTOR CLONES, IN CONTRAST, WERE RAISED IN AN ENVIRONMENT THAT HEAVILY FAVORED SIMILARITY.

Rudy is more aggressive, more impulsive than the others; Mark is more patient and kind; Miller is more disciplined. Ira is arguably the most dramatic of the Castors, but this may have been a learned behavior from the likewise dramatic Susan Duncan rather than an intrinsic trait. In any case, the Castors support the theory that both nature and nurture are important elements in forming personality and behavior. Despite being genetically identical and raised in an identical environment, the Castor clones, although more similar to each other than the Leda clones, nonetheless exhibit individual personalities.

Speaking of Castor, unlike seasons one and two of *Orphan Black*, season three does not look to scientists and scientific works for

its episode titles. Instead, the source for season three's titles is U.S. President Dwight Eisenhower's farewell address from 1961. While this may seem like a bizarre option compared to Charles Darwin or Francis Bacon, many of the themes President Eisenhower touched upon are relevant to *Orphan Black*, especially given season three's focus on the military and Project Castor. A big point in Eisenhower's address was the military-industrial complex — the country's military and its related industry in producing military materials. Eisenhower discussed the role of the country's military-industrial complex in the context of the political state of the world, believing that military power was necessary. When considering the military branch of the clone project, it becomes a bit terrifying to think of them as a necessary power, especially with the Castor arm of the clone disease doubling as a sterilization method.

Eisenhower also discussed the upswing of scientific research in his address. He spoke of the need to hold scientific research in respect, but he also warned against allowing policy to help create the scientific elite. Eisenhower encouraged the people and the government to work to balance technological advances with old ideas and ways. This struggle with scientific advancement is reflected in the world of *Orphan Black* as a whole with the clones trying to fit into a normal world despite their involvement with the scientific elite of Dyad. In season three, specifically, the Leda clones struggle to balance the forces of Project Castor and Dyad while uncovering the truth and keeping themselves safe.

The shift in episode titles from scientific work to the military-industrial complex parallels Project Castor's shift from scientific experimentation in pursuit of knowledge to scientific experimentation in pursuit of military power. Project Castor had its beginning at Dyad in the name of pure scientific research, but as soon as the military took over and separated the two projects, Project Castor was no longer rooted in basic science. All efforts of those associated with Project Castor are military-based, as is reflected in season three's episode titles.

NATURE, NURTURE, AND . . . EPIGENETICS?

The nature versus nurture debate doesn't *quite* cut it when it comes to explaining all of the differences between the clones. That's when we look toward another factor, epigenetics, which is sort of a meshing point between nature and nurture, our genetic code and outside factors that influence it.

COSIMA
The epigenetic implications
themselves are just *mind-blowing*.

(3.10 "History Yet to Be Written")

Epigenetics refers to the study of heritable changes in phenotype — traits that we can see — without changes to genotype, as well as the modifications of gene activity that are not based on alterations of DNA sequence. In simpler terms, epigenetics involves changes in gene expression that are not associated with the DNA sequence itself but involve chromosomal modifications that regulate gene expression, which can be either inherited or introduced later. You can think about the genome — DNA — as written in ink and the epigenome as penciled-in notes, transcribed right on top of the ink or added in the margins. The ink is unchanged, but the pencil might alter how the ink is read or interpreted. If you're musically inclined, another great analogy for understanding epigenetics is that of a piano score. DNA is the basic notation: the music notes and rests that map out the melody. Epigenetics are the articulations and dynamics: they tell the musician *how* the notes should be played — where the keys should be pressed more softly, where they should ring out loudly, and what sorts of nuances should be expressed. Change the dynamics and articulations and the melody can sound very, very different, even though the notes themselves do not change.

To further understand epigenetics, consider the cells in any person. Almost every cell in the human body contains a nucleus that holds all of the chromosomes, or genetic material, yet the human body is composed of dozens of different cell types — neurons, skin cells, liver cells, blood cells, and so many more. How does each cell know to express the genes specific to that cell type and repress genes

CLONE CLUB Q&A
Did they already know about epigenetics when Project Leda began?

No, not really.

The term "epigenetics" has been around for a while, but it did not become a field of study until pretty recently, around the year 2000. In the 1970s and '80s, there was no one studying, or theorizing about, epigenetics. Which is pretty interesting in terms of Project Leda. Epigenetics may explain why Sarah and Helena are fertile, and it wouldn't have even been on Dyad's radar as a possible issue.

of other cell types, when every cell has the same genome, the same DNA? The answer is within these epigenetic modifications.

Epigenetic markers are a group of chromosome modifications — chemical groups such as acetyl groups, ubiquitin groups, or methyl groups — that are added to either the DNA or the associated proteins of the chromosomes. These markers then determine if the associated gene is activated or repressed. One of the most common epigenetic markers on the DNA itself is methylation — the addition of methyl groups directly to the bases of the DNA — which is most often used as a marker to silence expression of a gene, and can function to block binding of transcriptional proteins. Large regions of DNA methylation are characteristic of regulatory elements that are shut off in the genome — a characteristic known as hypermethylation. When a region of the genome is hypermethylated, that means it is shut down, and the genes are not being transcribed or expressed.

EPIGENETICS COULD EXPLAIN WHY SARAH AND HELENA DO NOT PRODUCE THE INFERTILITY PROTEIN.

Some epigenetic information is inherited from our parents. When mitosis (cell division) occurs and DNA is duplicated, the chromosomal modifications are also duplicated to ensure that the expression patterns of the genes are maintained. This also happens in the gametes (reproductive cells) during meiosis, which is the reproductive cell version of mitosis. The epigenetics of the parents are passed down to the child.

However, you can also acquire new epigenetic modifications that are not inherited. Cellular interactions, foreign materials, the microenvironment of the cell, the activities of the person, and many other environmental aspects can cause alterations in the epigenome of a person. The womb is considered to be an important influence on the epigenome, right down to the minute microenvironments within it. Even identical twins can have epigenetic differences because they experience different microenvironments in the shared womb.

As for the clones of *Orphan Black*, who all share the same genome, epigenetics can explain differences between them. Some clones have the clone disease while others still haven't shown any symptoms. This may be because of epigenetics — they all have the gene for the disease, but some factor is causing it to be expressed in some clones and not others. In the same vein, epigenetics could explain why Sarah and Helena do not produce the infertility protein and therefore will not get the disease and *can* have children. It might even explain in part why Tony is trans and Cosima is gay (the biology behind gender and sexuality is still widely unknown, but there are now scientists looking toward the womb environment as a possible factor).

X-INACTIVATION

In humans, there are two main variations of sex chromosomes: XX and XY. (There are also intersex people whose sex chromosome composition is different from the two common types, but that is beyond the scope of this discussion.) The X chromosome is a full-length chromosome encoding numerous genes, and all people have at least one copy of it. The Y chromosome is a small, almost partial, chromosome that encodes for only some of the same genes as the X chromosome, as well as a few unique genes.

For all chromosomes other than the sex chromosomes, humans have at least two copies, and they are roughly identical: they encode for all the same genes unless there are deletions, but they may have different alleles (or variant forms) for the genes (such as an allele for type A blood from one parent and an allele for type O blood from the other parent). For every gene other than those on the sex chromosomes, we express two copies: the allele from one parent and the allele from the other parent.

People with XY chromosomes overall only have one copy of all the genes on the X chromosome. Therefore, they are only expressing a single dose of these genes. To equal out the dosage discrepancy for people with XX chromosomes, one X chromosome in every cell gets inactivated: only one copy of every gene on that chromosome is expressed in the cell. The inactivated chromosome gets turned into a tightly wound bunch of heterochromatin that can no longer be accessed for gene expression. This is known as the Barr body.

The pattern of X-inactivation is completely random, so not

every cell in the body has the same chromosome inactivated. In most people, the result is about a 50/50 split, resulting in roughly equal expression of the alleles of each gene from both parents. A good visual for this process is coat color on cats. The gene for coat color is on the X chromosome, and if the cat has an allele for black fur from one parent and an allele for white fur from the other parent, then some cells will cause the fur to be black and others will cause the fur to be white, depending on which chromosome is inactivated in each cell. This results in a patchwork coat, with areas of black and areas of white fur.

The Leda clones all started out with the same two X chromosomes, but during development they began to inactivate in a random process. No two clones are going to inactivate the X chromosomes in the exact same pattern. Now, for any genes in which the clones have the same allele on both chromosomes, this won't make a difference; but if there are two different alleles for any gene, the clones will vary in their expression of these alleles. This would make them not exactly identical, as far as gene expression is concerned.

There are thousands of genes encoded on the X chromosome, not all of which are known (many that are known are involved in diseases such as Duchenne muscular dystrophy, hemophilia, fragile X syndrome, and androgen insensitivity syndrome), so it's hard to say what exactly would vary among the clones due to differences in X-inactivation. But it is safe to say that this is another area of their genetics (along with their epigenome) that will differ among each of them. Just another example of how the genome doesn't have the last say in the final product.

THE OTHER GENOMES

Many people know from high-school science class that chromosomes in a cell's nucleus contain DNA, but many don't realize that nuclear DNA doesn't represent 100% of a person's genetic information. There are two other key pools of DNA that contribute to a person's genetics: the mitochondrial genome and the microbiome.

Mitochondria are organelles within a cell that are responsible for generating the majority of energy a cell uses to carry out its functions; in school, the first thing you learn about mitochondria is that they are the "powerhouses of the cell." Aside from the proteins and structures necessary for energy production, within the mitochondria are small pieces of circular DNA that contain roughly 40 different

CLONE CLUB Q&A

If all the clones have slightly different genomes, couldn't their synthetic sequences be detected by looking at variations in the clones' DNA? By directly comparing Alison's and Cosima's DNA to Sarah's, we should be able to see where Sarah's is different, right?

All of the clones have the same genome — that's what makes them clones! There is only one difference that was put into each clone on purpose, and that's the ID tag sequence. Dyad knows exactly where this sequence is located, so they can easily find it to identify a DNA sample and match it to a clone (however, Helena and Sarah, being identical twins as well as clones, have the same ID tag). Good thing they know where it is because the human genome is huge — about three billion base pairs long. If you imagine a base pair as a single letter in the code that writes your DNA, finding a tiny difference between two genomes is like trying to find a single typo in a 200,000-page book by comparing it to another 200,000-page book.

The first project to sequence all three billion base pairs in the human genome — aptly named the Human Genome Project — started in 1990 and was completed 13 years later in 2003. Sequencing a full genome is getting faster and less expensive all the time. There are even commercial services that offer to sequence your genome for you in as little as ten days. But comparing genomes from two separate clones would still be a very long process, especially if the person doing the sequencing didn't know where in the genome to look.

genes that encode for proteins used within the mitochondria. It may seem odd that an organelle has its own genome, but scientists believe the explanation for this is that millions of years ago, when Earth was nothing more than the proverbial "primordial soup," and life consisted only of single-celled organisms, one cell managed to engulf another cell, creating a team that worked together as a more successful organism. This led to the first cell with organelles — the eukaryotic cell that eventually gave rise to all multicellular organisms. Mitochondria contain their own genome due to their evolutionary beginnings as a single-celled organism.

An interesting attribute of the mitochondrial genome is that it is only inherited maternally. While the mother's egg appears relatively similar to other cells except for the fact that it only contains half the necessary chromosomes, the father's sperm is a very reduced cell, the cell body of which is made up of mostly the nucleus and no major organelles. When the egg and sperm fuse, the resulting fertilized egg has a genome with equal contribution from each parent, but the organelles within the cell, including the mitochondria, come only from the mother's egg.

There's an interesting implication when it comes to cloning: while all the clones have the same chromosomes as Kendall, they do not have the same mitochondrial genome as that would've required getting egg cells for each clone from Kendall's biological mother. Considering the large amount of Leda and Castor clones, it may be the case that they all have different mitochondrial genomes, due to the difficulty of obtaining that many eggs from a single woman. Even though the mitochondrial genome only contains a few genes, which only give rise to proteins used within the mitochondria, it is possible that differences in mitochondrial genomes between clones could explain some phenotypic differences.

The mitochondrial genome, because it has genes just like the nuclear genome, can contain mutations that result in mitochondrial diseases. Because mitochondrial diseases affect an organelle that is found in pretty much every cell in the body, they cause problems in multiple organ systems and can be hard to diagnose and treat. Most often, the most drastic symptoms are seen in the muscles, because muscle cells have high energy demands and therefore contain high numbers of mitochondria.

For women with known mitochondrial diseases who don't want to pass them on to their children, there's a fertilization process known as "three-parent babies," which prevents the inheritance of diseased mitochondria. Since the mitochondria in a developing embryo normally comes from the mother's ova mitochondria, this procedure instead uses the sperm from the father, the nuclear genome from the mother, and the mitochondria from an egg donor. This process is similar to SCNT cloning in that the nucleus of the mother's ova is taken and inserted into the donor's egg cell and then fertilized by the father's sperm. In this way, three people contribute genetic information to one embryo, hence the term "three-parent babies."

One key way we know all the clones differ genetically is the synthetic ID tags that have been inserted into their nuclear genome. These synthetically developed sequences unique to each clone allow the scientists at Dyad to identify the clones by their genome; if an unlabeled vial of clone blood were given to the scientists, they would be able to sequence the sample and know which clone it came from simply from the ID tag.

COSIMA

[...] they can tell us apart by
our DNA. That's how they knew you
weren't Beth.

SARAH

So we're not genetically identical?

COSIMA

We should be, but there's a
synthetic sequence.

(1.09 "Unconscious Selection")

In order to have the ID sequence function this way, the scientists need to know the location of the tag in order to sequence it, either via determining the exact chromosomal location (which is something that is nearly impossible to do in the laboratory) or via localization of this tag to a specific protein's sequence (which is a routine laboratory method). By connecting the ID tag sequence to a protein sequence, the ID will be included when the scientists sequence the protein, making it simple to find and compare it to the sequences of the other clones.

The gene sequence that the Dyad scientists chose to anchor the ID tag on is the sequence for cytochrome c, a protein that's part of the electron transport chain in mitochondria. An interesting note about cytochrome c: it's a highly conserved protein, meaning that the protein structure and sequence remain the same or similar through evolution across all species, plants, animals, and unicellular organisms alike. This high level of evolutionary conservation makes it a great tool for studying genetic changes across species and evolutionary relationships between species. It may not be a coincidence that a gene so key to evolutionary biology was chosen by Dyad to contain the ID sequences of the next step in human evolution.

Aside from the nuclear genome and the mitochondrial genome, recently scientists have been considering what is sometimes referred to as the third genome: the microbiome. The millions of bacteria found in and on a person's body are mostly composed of symbionts within the digestive tract; one study suggests that more than *five thousand* different types of bacterial groups may make their home

---

CLONE CLUB Q&A
Why are Cosima's boobs bigger?

Although unconfirmed as fact, in season one, Felix pointed out that Cosima's boobs seemed to be bigger than the other clones'. He has an artist's eye, so he notices these kinds of things.

Breast size is genetic, but it can be influenced by diet and environment during early puberty, which could explain differences between the clones. Or Cosima might just have a very nice collection of push-up bras.

in the gut. While these bacteria and their genetic information are not inherently part of a person, they are necessary for our survival. For the longest time, people have demonized these bacteria, but evidence shows that our microbiome aids in digestion, supports the immune system, and may have implications in a number of diseases as well as in how our bodies interact with drug treatments. It's only been in recent years that we've seen widespread interest in boosting microbial health in our guts, for example by eating prebiotic and probiotic foods. The microbiome, then, is one of the factors that researchers have been considering as trends move toward personalized medicines, drug treatments that are optimized for the individual rather than blanket treatments that target the disease and ignore the genetic and biological factors that affect how a person's body interacts with the treatment. These microbes also contribute to genetic variation; they introduce gene functions that our human cells then do not have to evolve on their own. The composition of these critical bacteria is determined mostly by our environment — the places we live, the food we eat, what we expose ourselves to on a regular basis — making the microbiome of a person a unique feature.

Just as the *Orphan Black* clones' fingerprints differ, the levels and types of microbes that live on and in their bodies would differ, too: each clone has an individual microbiome caused by environmental differences between them, and these microbiomes could influence differences in health. For example, since the microbiome plays a role in immune system function, it is possible that differences between the clones' microbiomes could lead to differences in how each of their bodies responds to the clone disease, which could explain in part why some of the clones have yet to start showing any symptoms, even though we know that they are not immune to the disease.

WHY DOES COSIMA WEAR GLASSES?

Personality traits and behaviors don't account for all the differences between the clones; some are physical. One of the most obvious examples is Cosima, the only known glasses-wearer within the Leda

clones. But why does she need glasses, especially since none of the other clones seem to have any problems with their eyesight? The answer is actually quite simple: Cosima needs glasses because she's such a nerd.

That may seem like a joke, but it's actually true! Cosima needs glasses because she spent so much time during her childhood reading and studying (probably for many hours at a time in less than ideal light conditions), and over time that strain on her eyes caused her vision to worsen. This led to a physical difference in Cosima from the other clones, a direct result of her environment, her habits, and her life experiences. Perhaps all the Leda clones have a propensity for their vision to worsen over time, but Cosima is the only clone we know of whose lifestyle has played into this risk and caused a physical outcome. A very interesting example of nature versus nurture.

The clones are different and individual, there's no arguing that.

There are too many factors that contribute to personal development — physical, cognitive, and behavioral — to realistically expect the clones to be perfect carbon copies of each other, even if they had been born and raised in exactly identical vacuums. Even Cosima, as fascinated as she is with the biological and genetic determinants that make people who they are, would acknowledge that taking one side of the nature versus nurture debate, calling herself either a determinist or a behaviorist, would be completely impractical. Choosing sides leads to a single-cause fallacy. There are a surprising number of people who still subscribe to the belief that one factor dominates the other, or even just think that "nature" and "nurture" are the only two factors. The truth is, we don't fully understand the scope of what makes us who we are. Isn't it more interesting to think that creatures as complex as human beings are each unique thanks to so many factors?

CLONE CLUB Q&A
What's the point of having monitors? If Neolution's goal was to create the Leda clones as proofs of concept, why didn't they just keep the clones at the lab?

As far as we know, Rachel is the only Leda clone who was raised self-aware (and even then, she was only self-aware from the age of six). All the other Leda clones that didn't escape Dyad's purview were placed with families and raised naive to their status as clones. They were assigned monitors — either family members or friends or romantic partners — who reported their health and behaviors back to Dyad.

What's interesting is that the monitors were also naive. In some rare cases, like with Donnie Hendrix, the monitors were already in an intimate relationship with the subject. Fascinatingly, according to Donnie, he agreed to be Alison's monitor simply because his sociology professor asked him to do so as part of a longitudinal study, telling Donnie that doing

so would contribute to science. For whatever reason, he didn't question this and kept up with reporting data a good decade or so later. Most of the monitors, like Beth's monitor, Paul, and Cosima's monitors, Emi and Delphine, were given their assignments first and then instructed to engage the subjects. But other than Delphine, who worked actively on the clone projects, they didn't know that their subjects were clones.

Cosima guesses correctly in "Conditions of Existence" that they are part of some sort of double-blind experiment. An experiment is said to be double-blind when both the tester and the subject are missing important information about the study until after the outcomes are known. The point of this is to prevent any bias, even unconscious bias that comes with knowing, that might skew the study's results. The interesting thing here is that it's unclear exactly what Dyad is measuring and how the different monitors are reporting on their subjects. Even if they were asked to report on the same things, Delphine, a scientist, would certainly report differently than, say, Donnie. It makes for a pretty inconsistent study on Dyad's part.

Cosima also mentions that if a blinded subject were to become self-aware, it would be best to terminate the experiment but, despite a number of clones now being self-aware, this hasn't happened. In fact, Alison and Donnie have even continued to diligently provide data, for whatever reason (but they seem to be the only ones doing so). In the interest of collecting data and advancing knowledge, Dyad and Neolution seem interested in learning as much as they can about the clones' development in different environments, even if this learning is secondary to the reason the clones were created in the first place.

CASE STUDY:
COSIMA NIEHAUS

ID: 324B21
DOB: March 9, 1984
Birthplace: San Francisco, California,
United States
Status: Alive

COSIMA

I am going to get you so baked one day.

(1.06 "Variations Under Domestication")

Throughout *Orphan Black*, we witness the resident geek monkey smoking her fair share of weed — she even starts growing it in her underground lab in season four. Although Cosima starts the series as a pleasure smoker, she eventually uses the marijuana that she grows to mitigate the symptoms of the clone disease. Many peple living with chronic illnesses use marijuana to help ease pain, and the treatment is spreading as legalization progresses across the United States and Canada.

COSIMA

We have a little crop right there.
Totally organic.

(4.02 "Transgressive Border Crossing")

Marijuana's key chemicals of interest for medicinal purposes are cannabinoids, the most well known of which is THC (delta-9-tetra-hydrocannabinol). THC has mind-altering properties, which is of interest to recreational users of marijuana, as well as therapeutic properties, such as the ability to reduce pain, nausea, and inflammation, and the ability to increase appetite. Another cannabinoid found in marijuana, cannabidiol, can be used to treat seizures, pain, and inflammation and does not affect the user's state of mind. There is no hard and fast definition for medical-grade marijuana, and anyone using marijuana for healing purposes can be considered to be using therapeutic marijuana. Typically, therapeutic marijuana has a high potency and is organically grown, not mixed with any other chemicals or drugs.

Smoking marijuana allows for the quickest absorption of cannabinoids. The inhaled smoke enters the lungs and the chemicals then flow into the bloodstream, where they are quickly transported to other organs of the body, including the brain. Eating cannabinoids allows for slower absorption but is still effective. Once the THC is in the brain, it acts on various receptors, causing the user to feel "high" as well as experience the various therapeutic effects. The chemicals bind to receptors on the neurons in the brain, causing these neurons to fire — that is, release chemicals that further interact with cells throughout the brain — triggering a feeling of euphoria. The neurons continue to fire and maintain the high until the cannabinoids are cleared and the receptors are freed, ending the neuronal response.

Clinically, the most successful therapeutic use of marijuana is for nausea relief. Medical marijuana is a common treatment for chemotherapy patients and people with late-stage AIDS. Currently, there are medications on the market that are cannabinoids in pill form, commonly prescribed to cancer patients. However, this treatment is relatively short-term and requires frequent repetitive use to be effective. Clinical trials have been held to test the effectiveness of THC on treating nerve pain and multiple sclerosis. While some results have been promising, there remain no drugs on the market containing THC, other than nausea relief pills.

Despite the lack of options available on the market, research has vastly expanded in terms of diseases that cannabinoids might effectively treat. Epilepsy, Crohn's disease, and cancer are the most promising candidates for cannabinoid treatment, but research covers a plethora of diseases, including Alzheimer's disease, PTSD, lupus, and glaucoma. As legalization of marijuana spreads, research, clinical trials, and commercially available medications will continue to expand and produce results.

For Cosima, aside from the pleasures she has always enjoyed, marijuana helps ease the nausea and pain associated with the clone disease. Although her worst symptoms include lung polyps and coughing up blood, smoking is probably most effective for her due to the speed at which the chemicals enter her bloodstream and begin to ease her pain. The lobes of the lungs are made up of little sacs, known as alveolar sacs, where gas exchange takes place with every breath. At these locations, the alveolar sacs are directly next to blood vessels, which facilitates efficiency in the gas exchange. This proximity to the blood vessels is why cannabinoids can enter the bloodstream so quickly. If Cosima were to ingest marijuana (by baking it into a brownie, for example), the cannabinoids would have to travel down the esophagus, through the stomach, where the food and drug would be digested, and then into the intestines, where they could finally be taken up by the bloodstream.

Although Cosima partakes in this therapy illegally, the benefits are worth the risk; her disease is at the stage where even the smallest bit of relief seems monumental. After all, as she discusses with Mrs. S in season four's "Transgressive Border Crossing," she can no longer use Kira's bone marrow; while testing the vectors for the gene therapy, bone marrow transplants would alter any potential effects of the tests, and therefore interfere with the results. Cosima also mentions that the weed helps her with her appetite, due to the cannabinoids acting to promote hunger. (Recreational marijuana users refer to this as "the munchies.") A good appetite is necessary for keeping Cosima's strength and health at their highest standards while she is stem cell treatment–free and testing therapies. Therefore, her only means of managing her symptoms and keeping her in the best state is through the use of her private stash of medical marijuana.

CLONE CLUB Q&A
Why is helium way funnier than polonium?

In one memorable scene from season two's "Variable and Full of Perturbation," Cosima and Delphine are having more than a little bit of fun with weed and helium balloons when Cosima declares "I am helium, and I'm way funnier than polonium!"

While this seems like a nonsensical stoner pronouncement, it's actually a reference to the fact that tobacco leaves, the main ingredient in cigarettes, contain the chemical element polonium (which, by the way, is super radioactive and causes lung cancer). Basically, Cosima is saying that getting high is way more fun than smoking cigarettes (a habit that we know from season one Delphine indulges in). No matter what state she is in, Cosima remains a huge science nerd.

CHAPTER 3

"YOU'RE JUST A BAD COPY OF ME"

KENDALL MALONE: ONE PERSON, TWO CELL LINES

KENDALL MALONE
Hell of a twist of nature, isn't it?

(3.09 "Insolvent Phantom of Tomorrow")

Some eight years after the Leda clones were created, the proverbial mold was thought to be broken: the Duncans' work was lost in a lab fire. Nearly everything, including data from the clone experiments and copies of the original Leda genome, was destroyed. The research teams were unable to replicate the cloning experiment for over a decade before they earned a single success in Charlotte. When the clones began to fall ill with what appeared to be a genetic disease, there was no record of the original genome, or of the synthetic sequences added to the clones' genomes, to consult. From what we learn of the failed treatments Dyad and others have attempted, it's clear that finding the original genome is pretty much integral to locating the infertility sequence and finding a cure for the clone disease.

In season two, Ethan Duncan, the clone creator long thought to have died in the fire that destroyed his lab, is found alive and in

possession of what might be the last digital copy of the Leda genome. It's in an old format — computer nerd Scott can't get his hands on a 6502 processor to read the file (this particular processor was produced in 1975, around the time Project Leda was just beginning, and it's one of the technologies that eventually led to the development of modern home computers). But he does find a system that shares a chipset with the similar 65C02 and is thankfully able to handle the data . . . and, of course, said data is encoded. To encode and protect the genome, Ethan Duncan used a Vigenère cipher, which is what's known as a polyalphabetic code. This form of cipher, which has been around since the 16th century, uses two or more cipher alphabets to encode text. To solve it, it's not just a matter of figuring out that S means A and B means T, for example, and solving for each letter one by one. In a Vigenère cipher, the letters shift by different amounts, using a word or phrase as the encryption key, and so more than one letter can be decoded to reveal the same letter; that is, an S and a B might both reveal an A, but S might also reveal a T later in the code, depending on the encryption key. This type of encryption is especially useful for encoding a genome, which is a series of four letters: A, T, G, and C. A monoalphabetic form of encryption would simply be too easy to solve. Today, the method for a Vigenère cipher is well understood, but for nearly three centuries, it held up as such an effective code that it was known as *le chiffre indéchiffrable*, the undecipherable cipher. Ethan reveals in "Things Which Have Never Yet Been Done" that the cipher encodes the infertility sequence and that he has brought along the key. All Scott has to do is design an algorithm using the key to transcribe the encoded infertility sequence into its original genomic code. The rest of the genome remains encoded, each sequence requiring a separate and distinct key that Ethan refuses to provide. To unlock its secrets, Cosima and Scott would have to solve the key for each sequence and transcribe each independently, which would take too much time, especially with Cosima's health continuing to decline. Frustratingly, Ethan dies before they can convince him to help them decode the rest of the genome. We're more than curious to know what key words or phrases Ethan might have selected to encode the clones' various genetic sequences.

In the season two finale, Kira reveals to Cosima that the copy of H.G. Wells's *The Island of Doctor Moreau* that Ethan Duncan gave her is filled with a symbolic code, along with notes and illustrations (we discuss what we can see on the pages and what it means in chapter six), and Cosima believes that this code must be the encryption key to the rest of the Leda genome. Duncan's book choice is interesting: *The Island of Doctor Moreau* deals with themes of humans "playing God" with nature, human identity, and bioethics. It also features animal-human hybrids created through vivisection and physical splicing rather than by genetic manipulation.

It turns out that Rachel is the only surviving person who knows how to decode the book. Scott realizes this when he recognizes symbols in Rachel's paintings that closely resemble symbols in the book. Ethan Duncan, we learn, taught Rachel the code as a sort of game when she was a child. In "Ruthless in Purpose, and Insidious in Method," Rachel translates one page from *The Island of Doctor Moreau* to reveal a nursery rhyme: "In London town, we all fell down, and Castor woke from slumber. Find the first, the beast, the cursed. The original has a number." The rhyme is accompanied by a cryptic sequence: H46239.

And so, in "Insolvent Phantom of Tomorrow," Sarah, Felix, and Mrs. S travel to London, England, to track down the source of the sequence, which one of Mrs. S's contacts in London, Terry, reveals is an old prisoner number. They expect that this prisoner must be the Castor original, and that discovering this person will bring them one step closer to finding the Leda original. What they don't expect to find is Kendall Malone, Mrs. S's estranged mother (rightly so, it seems — Kendall was imprisoned for killing Siobhan's husband, John), a tough, chain-smoking woman who just wants to be left alone. Kendall Malone also just so happens to be the original to both Castor *and* Leda.

> **SARAH**
> Castor and Leda are ... are
> siblings, S.
>
> **MRS. S**
> So what?
>
> **SARAH**
> So if your mum has two cell lines,
> that means that she's our original,
> too. We need her. She's Leda.
>
> (3.09 *"Insolvent Phantom of Tomorrow"*)

By a quirk of biology, Kendall is able to be the genetic original to both clone cell lines. She is what is known as a chimera. In Homer's *The Iliad*, the Chimera was a "grim monster sprung of the gods," a three-headed, fire-breathing hybrid creature, part lion, part snake, and part goat. In medical terms, a chimera is a hybrid of a different sort: a person who has populations of cells derived from two independent (human) sources. The most common form of chimerism occurs in the case of dizygotic, or fraternal, twins. The twins start off as two separate eggs fertilized by two separate sperm growing together in the womb. During development, one twin (in this case

Kendall) will absorb the other twin (in this case her brother), a process often referred to as "vanishing twin syndrome." There were two fertilized eggs, but then suddenly there's only one! But the cells from the vanished twin haven't really vanished at all. These cells become integrated into the host twin's developing body, where they may be localized to a certain organ or tissue, like a kidney or specific patches of skin, or distributed throughout the body, creating a patchwork of two genomes in the person. This most common form is often referred to as tetragametic chimerism because the cell lines are caused by two separate sperm that have fertilized two (initially) separate eggs. (Chimeras are also referred to as tetragametes because their cells are from four genetic sources: two eggs and two sperm cells.)

There are also smaller instances of chimerism, known as microchimerism, where the concentration of cells from the second cell line is so low that it might be barely detectable. This can occur through organ transplants or blood transfusions, or through twin pregnancies where the developing fetuses exchange blood. There have also been recorded instances of microchimerism in mothers from the children that they've carried, especially when the fetus is lost in utero. These microchimerisms might be transient and fleeting, or they might last years, decades, or a lifetime.

Two or more populations of cells with distinct genotypes can occur in people without their having absorbed a twin egg. That is, some people can develop distinct cell populations that are all derived from a single fertilized egg. This is known as genetic mosaicism, a phenomenon that should not be confused with chimerism. Most forms of mosaicism in humans are tied to trisomies, which are the result of errors in meiosis, the cell division process that forms eggs and sperm. Reproductive cells only have one set of chromosomes (whereas the rest of the cells in the body have two sets). Trisomy occurs when the chromosome pairs fail to properly separate during cell division (an error known as chromosome nondisjunction), and the resulting cell ends up with an extra set of one of the chromosomes. The most well-known trisomy is trisomy 21, or Down syndrome, where an individual has a third copy of chromosome 21.

BY A QUIRK OF BIOLOGY, KENDALL IS ABLE TO BE THE GENETIC ORIGINAL TO BOTH CLONE CELL LINES.

Having a trisomy does not automatically mean that you have mosaicism. Mosaicism only occurs if the trisomy is present in a selection of the body's cells, but not all of them. (Most trisomies affect all cells.) This can happen in one form of Klinefelter syndrome, 46/47 XY/XXY, where some cells have the standard 46 chromosomes, while other cells have an extra X chromosome, totaling 47. Another sex chromosome (XY-linked) disorder that sometimes shows mosaicism is Turner syndrome, where some cells might have both X chromosomes, while others will demonstrate monosomy, or only one X chromosome. In general, conditions like these that show mosaicism, rather than a complete genotype for the disorder, tend to show milder symptoms because it's not present in all cells.

Mosaicism can also occur in somatic cells (non-reproductive cells) and can be caused by a gene mutation during fetal development that affects the genotype of some but not all cells in the body. In this case, the mechanism that keeps one of each chromosomal pair turned off works incorrectly, resulting in different chromosomes being turned on in different cells. This kind of mosaicism is fairly commonly found in nature. If you've ever seen a tortoiseshell cat, the multicolored pattern of its fur is an example of mosaicism in action.

It's thought that many, if not most, cases of human chimerism go unidentified. In a case where the absorbed sibling is of the same sex, it is less likely that it will be noticed at all. There would be no reason to suspect that a person might be a chimera, unless the chimera's second cell line forms a distinctive tissue or organ (such as in the case of intersex chimeras, who, along with their own sexual organs, might also develop sexual organs from cells of their sibling of the opposite sex; there has been at least one case of a chimera being treated for complaints of an undescended testis only to find upon examination that he possessed an ovary and a fallopian tube formed by his vanished sister's cells), or unless that second cell line causes medical issues. If the chimerism is between siblings of different sexes, it can sometimes lead to hormone problems during growth and development, depending on which organs are affected. In the case of Kendall, her reproductive organs were likely not influenced by her brother's cells given that she was able to successfully have a child: her daughter, Siobhan Sadler, a.k.a. Mrs. S.

Sometimes chimeric traits *are* visible. Heterochromia, which describes a difference in coloration of the pigments in the iris (such as when a person has one brown eye and one blue, or a distinct pie slice of a different color in their iris) or skin or hair, *could* be a visible indicator

> IT'S THOUGHT THAT MANY, IF NOT MOST, CASES OF HUMAN CHIMERISM GO UNIDENTIFIED. IN A CASE WHERE THE ABSORBED SIBLING IS OF THE SAME SEX, IT IS LESS LIKELY THAT IT WILL BE NOTICED AT ALL.

of a second cell line. That said, there are many factors that can cause heterochromia, running the gamut from a genetic mutation to disease to injury — even a penetrating injury to the eye by an iron-containing object can deposit iron in the eye and alter eye color.

If a chimeric person's cell lines are both present in their skin, then their chimerism might be visible along what are known as Blaschko's lines. Skin cells grow in a very specific pattern across the body, and this pattern is common for everyone: a V shape down the middle of the back, S-shaped swirls across the chest and sides. The pattern is created as embryonic tissue divides and stretches to form skin. In a chimera with two cell lines working at the same time to form skin, one cell line pushes and swirls through the other like steamed milk poured into a shot of espresso. Most people have genetically identical skin cells, so these lines are invisible; however, there are subtle differences between the cells of various areas of skin. So even a non-chimeric person who develops a skin condition that presents as a rash or change in pigmentation might reveal the patterns as the infection or disorder follows Blaschko's lines. Chimerism might be visible as distinct pigmentation differences along Blaschko's lines, and subtler differences in pigmentation, too difficult to see with the naked eye, might be revealed using a UV light.

The chimera immune system is interesting: it has incorporated a second, technically foreign, cell line and is generally tolerant of it. Think about blood groups: ABO groups are defined by what antigens (if any) are present on the surface of red blood cells and what antibodies are present in the plasma. (Quick refresher: antigens are molecules that can trigger immune responses in host bodies, and antibodies are proteins that are produced by the body to neutralize those non-self antigen-marked agents.) We know that it's important to consider blood groups for transfusions because antigens different from the host's can trigger a dangerous immune reaction: a person with type A blood produces anti-B antibodies in their plasma, so you can see where things might go horribly wrong if they were to receive type B blood. That's where chimeras are fascinating: they can tolerate non-self cells without triggering an immune response. There is even a classification of chimerism, known as blood group chimerism, where an individual can have red blood cells from more than one group coursing through their veins at the same time — even blood groups that should be incompatible. This is more common than you might think. Twins who share a placenta, known as monochorionic twins, mix blood systems where they share placental blood vessels, thus mixing hematopoietic, or blood-making, stem cells. This can

CLONE CLUB Q&A
Would Leda/Castor be possible if the original donor had Klinefelter syndrome?

Klinefelter syndrome (also known as XXY syndrome) occurs when a person possesses a Y chromosome and two or more X chromosomes. Humans usually have 46 chromosomes, but this condition results in a person having 47 or more. Typically, a person with Klinefelter syndrome has subtle symptoms, such as taller adult height, weak muscles, poor coordination, and less-developed sex characteristics. Most significantly, however, a person with Klinefelter syndrome is infertile.

Technically, a person with Klinefelter syndrome has the genetic material necessary to make genetically male (XY) and genetically female (XX) clones. But a person with Klinefelter syndrome still only possesses one cell line, whereas Kendall Malone, the Leda/Castor original, possessed two distinct and separate cell lines. A Klinefelter donor's sperm would have to be surgically removed and microdissected to get the genetic material needed.

It is possible to have a Klinefelter donor — *in vitro* fertilization treatments using microdissected Klinefelter sperm have resulted in pregnancy, but the success rate is limited, lower than 50%. Using genetic material from a person with Klinefelter syndrome would entail lot of extra work and risk on top of the high level of work and risk that already exists with cloning standard XX or XY cells. The failure rate would probably be higher.

happen with dizygotic (non-identical twins) as well as monozygotic twins. Cells, and sometimes stem cells, can be exchanged in amniotic fluid, too, if the twins also share an amniotic sac.

One of the causes for this immune tolerance might be that the inclusion of the non-self cells, whether from a vanished twin or from blood exchange in the womb, happens while the immune system is still forming. Because these foreign cells are there from pretty much the beginning, they get incorporated into the immune system as part of the cellular self. One advantage to this is that chimeras can have a wider tolerance for receiving blood and stem cell transfusions and organ transplants.

People who acquire cells from a second cell line *after* their immune systems have formed do not get to reap the same immune advantages as true chimeras. One common form of microchimerism, fetal-maternal chimerism, can be caused by cells, including stem cells and progenitor cells, transferred in the exchange of fluids between mother and fetus. These fetal cells have been found to migrate all over a mother's body, to become part of the heart or the brain, or to be found in joints or in the blood. Most fetal cells that make their way into the maternal body are eventually found and removed thanks to the immune system, but the undifferentiated cells might survive and proliferate in the maternal system. Studies have observed the presence of chimeric cells in the blood and tissues of healthy women long after pregnancy. Recent research has reported a link between these fetal cells and the post-pregnancy development of autoimmune disease in the mother.

Microchimerism is not uncommon, but not all cases of microchimerism develop into autoimmune disease (and most people experience no effects from their microchimerism). It's suggested that some agent triggers the hibernating fetal cells to attack maternal host cells and initiate autoimmune disease, but the jury's still out on what those triggers might be. Of course, in the case of fetal-maternal microchimerism, the exchange of cells happens in both directions; it's also been suggested that chimeric maternal cells found in the child's body, which can also persist for long after birth, might be implicated in some early childhood autoimmune disease.

Outside of any illness or visible physical traits caused by

chimerism, the condition would only become apparent with tissue sampling and DNA sequencing, and only if both cell lines are retrieved in the sampling process. This is exactly what happened with Kendall. Our guess is that the samples were a standard blood sample and cheek swab for epithelial cells. If Kendall Malone were a convict in the United Kingdom today, samples would have been taken as part of her arrest. Her entire genome would not be kept on file; rather, only sections known as short tandem repeats, or STRs, would be analyzed. For the most part, the human genome is identical across all humans, but these small sections of repeating sequences are variable from person to person. The samples also undergo a sex determination test, which, if the samples Kendall provided contained both cell lines, would have made it easier for Ethan Duncan to identify her chimerism.

The U.K. set up the first national DNA database, officially known as the U.K. National Criminal Intelligence DNA Database, in 1995 to have a record of samples from criminals and suspects as well as samples collected from crime scenes. This is similar to the national

DNA database CODIS (Combined DNA Index System) developed for the FBI and used in the United States and Canada. Back when Ethan Duncan would have been searching for the ideal donor for his human cloning project, however, techniques used for DNA profiling — the polymerase chain reaction (PCR) in the U.S. and DNA fingerprinting in the U.K. — were only just being finessed. DNA fingerprinting was only used for the first time as a criminal investigation tool in 1986, after the Leda clones had already been born. So instead of having access to a database of prisoner DNA, Ethan did rounds of testing inmates, both men and women, under the pretense of a cancer research project. He wasn't looking specifically for a chimera, just two promising cell lines, likely with gene markers suited to his concept of an ideal baseline. But someone like Ethan Duncan could never encounter something as biologically compelling as two cell lines contained within one individual and pass up on the opportunity to experiment.

The most famous case of chimerism is that of Lydia Fairchild. In 2002, Fairchild was separating from her long-term boyfriend and was applying for child support for her two children (and at the time she was also pregnant with her third child). She was required to prove that she was the biological mother and underwent routine DNA testing, supplying samples of her DNA and her children's DNA. The test results confirmed that her partner was the biological father of the children, but suggested that she was not the biological mother. Rather, her DNA was only similar enough for her to be a biological aunt to her own children. She was accused of being a fraud or of secretly having been a surrogate to her children, and she nearly lost custody of them. The court ordered a court officer to be present to witness the birth of her third child and for the child to be tested immediately at birth. That child was also found not to be genetically related enough to her to be hers.

At the time of Fairchild's case, human chimerism was known but had not been much explored. The early 2000s were apparently a big time for human chimeras. In 2002, 52-year-old Karen Keegan was also identified as a chimera. She had begun suffering from kidney failure a few years earlier, and her family underwent testing to see if they would be suitable kidney donors. The results of these tests revealed that Keegan could not possibly be the biological mother of two of her three sons. This case confounded doctors for a few years before they realized that she must be a chimera. Her blood cells carried different DNA than the cells in her ovaries.

It was coverage of this case that cued someone to test Lydia Fairchild's tissues for evidence of chimerism. They tested several cell types and found that, while her blood and several other tissues were not a match for her children's, cells in her thyroid and cervix were. Like Karen Keegan, Lydia's blood as well as at least some of her cheek cells were formed from her vanished twin's cell line, while her thyroid cells and her ovarian and cervical cells were her own. In this case, her children were genetically her own, but the inverse could also have been possible: in cases where reproductive organs are formed by the secondary cell line, it is possible for a chimera to give birth to children who are genetically their vanished sibling's.

Human chimeras are naturally occurring, but nonhuman chimeras have been created in laboratory settings. Laboratory-created chimeras tend to be interspecies chimeras, such as the "geep," a sheep-goat chimera first reported in 1984. An interspecies chimera shouldn't be confused with a hybrid; a hybrid would be created either by introducing sperm from one species to an egg from the other or by inserting genes from one species into the genome of the other. The geep, on the other hand, was formed by fusing a sheep embryo with a goat embryo, and it therefore has four genetic sources: one sheep egg and one sheep sperm as well as one goat egg and one goat sperm. It's a tetragamete. As would be expected, the geep demonstrated a patchwork of both goat and sheep tissue: areas of the body with sheep integumentary cells grew wool, and areas with goat cells grew hair. The possibility of fusing embryos of different species harkens back to the mythological version of the chimera and, as explored in *The Island of Doctor Moreau*, raises the possibility of human–nonhuman chimeras. *Doctor Moreau* was written well over a century ago, but

THE UNITED STATES AND THE UNITED KINGDOM HAVE YET TO ISSUE ANY STRICT REGULATIONS REGARDING CHIMERA RESEARCH.

this possibility is something that's recently been on more than a few people's minds. In fact, in January 2017, Jun Wu and colleagues published their success in creating pig-human hybrid fetuses. Before you start imagining Doctor Moreau–style creatures, know that the hybrids with the largest proportion of human cells were estimated by Wu to have only about one human cell for every 100,000 pig cells at most. The research team was inspired by shortages of human organs for transplantation to attempt a hybrid animal that could grow suitable human organs. The embryos were terminated after three to four weeks of development. As much as this is a landmark in chimera research, ethics regulations guide just how far these experiments can be pursued.

In 2004, the Canadian government issued the Assisted Human Reproduction Act to regulate research related to human reproduction. One of its provisions specifically prohibits the production of a chimera for any reason. As stated in Section 5 of the act: "No person shall knowingly [. . .] create a chimera, or transplant a chimera into either a human being or a non-human life form" or "create a hybrid for the purpose of reproduction, or transplant a hybrid into either a human being or a non-human life form." So, if Neolution had any plans for creating chimeras, their research would be all kinds of illegal at their Canadian base (although it's not as if Canada's similarly prohibitive stance on human cloning experiments has prevented Neolution in the past). The United States and the United Kingdom have yet to issue any strict regulations regarding chimera research, but in August 2016, the U.S. National Institutes of Health called for a lift on the moratorium on chimera research and proposed changes to the review processes for experiments involving them. Mind you, even with restrictions, it would be difficult to monitor whether regulations were being followed if research was being conducted illegally.

TWO SEPARATE CLONES FROM ONE PERSON: ISOLATING CELL LINES

It's amazing enough that Kendall has two separate genomes within her cells — but how did Susan and Ethan go about creating two separate groups of clones from her? The main hurdle would be isolating the two separate cell lines within Kendall: one to use as the Castor original and one to use as the Leda original.

Once the Duncans got hold of some of Kendall's biological material (most likely a blood sample or cheek swab), they needed to find a way to propagate the cells in a culture system so they would have enough material to work with while figuring out the cloning process. They would have needed to know which cells were XX (to become Leda) and which cells were XY (to become Castor) as early as possible in the process, so they didn't have to continually worry about the purity of their cell populations.

The simplest way to determine the genotype of a cell is to lyse the cell (break open the cell membrane to gain access to the contents) and use a polymerase chain reaction (PCR) to determine which cells have which genome. PCR is a laboratory technique that uses the enzyme that normally replicates DNA to make tens of thousands to millions of copies of a region of DNA. By amplifying a region of DNA to this extent, scientists can study a gene of interest from very little starting material. This amplification also allows for the DNA to be visualized on an agarose gel, putting regions of the genome into a form that can be visually compared to other DNA samples.

Simply, PCR begins with a small amount of template DNA (containing the region of interest), a pool of DNA nucleotides used to build new fragments of DNA, the DNA polymerase enzyme that replicates DNA from a template, and small synthetic pieces of DNA called primers that are used to guide the enzyme to the target sequence on the template DNA. When these components come together at the correct temperature, the result is a series of reactions that replicate the strands of DNA exponentially.

This DNA amplification technique can be utilized for a myriad of experiments, but it is very often combined with short tandem repeat (STR) analysis for genotyping, as mentioned earlier. STR analysis refers to the use of small regions of the genome (known as microsatellites) containing repetitive sequences that are unique to a person in order to distinguish between two DNA samples. For example, region X of the genome may be a microsatellite, and person A has 13 repeats at this position while person B has 24 repeats at this region. When region X is amplified for both people using PCR and visualized on a gel, the scientist will be able to distinguish the samples because person B will have a much larger sample in this region due to having 11 more repeats than person A.

STR analysis could have been used to identify the two different cell lines within Kendall. Even though Kendall's two genomes are biological siblings, they would still have unique microsatellites, allowing the scientists to distinguish the genomes of any of Kendall's cells tested in this manner. This is also how Scott figures out that the samples being used for stem cell treatment for Cosima in "Knowledge of Causes, and Secret Motion of Things" belong to Kira (based on the similarity of microsatellites between the two samples).

However, this technique requires killing the cell, which would not be helpful to scientists trying to get enough starting material for two separate cloning experiments.

THE XX CELLS COULD BE TAKEN TO PURSUE THE LEDA CLONING PROJECT, AND THE XY CELLS COULD BE USED FOR THE CASTOR CLONING PROJECT.

It seems the best choice for the Dyad scientists would have been to isolate cells from Kendall's samples into single cells as soon as they collected the sample. If single cells were then put into culture and allowed to grow, they would give rise to colonies of tens of thousands of cells that are all clones of each other, since they all originated from the division of a single starting cell. Once the cultures were large enough, some cells could be taken to genotype each colony, knowing that all the cells from the colony will be of the same genotype. Then the XX cells could be taken to pursue the Leda cloning project, and the XY cells could be used for the Castor cloning project.

CHARLOTTE

You're my big sister.

SARAH

How old are you?

CHARLOTTE

Eight.

SARAH

I have a daughter your age.

CHARLOTTE

Her name is Kira. I'm her cousin.

MARION

Aunt, actually. But we're going to go with "cousin" for now.

(2.10 "By Means Which Have Never Yet Been Tried")

Kendall's family tree is unique for a number of reasons, all stemming from her chimerism. Having two different genomes means that she was able to give rise to two lines of clones: Project Castor, all of whom share about 100% of their genome with the XY cells in Kendall (the reason for this being "about" 100% and not absolutely 100% is the variations in their mitochondrial genome, as discussed in chapter two), and Project Leda, all of whom share about 100% of their genome with the XX cells in Kendall. In addition, the Leda and Castor clones share roughly 50% of their genomes with each other, because Kendall's XY genome was from her absorbed sibling, making the Leda and Castor clones genetic siblings.

Of course, the Clone Club family tree extends beyond Kendall and the two groups of clones. Kendall had her own child, Mrs. S, and as is normal for a biological mother and child, they share roughly

50% of their DNA from Kendall's XX genome. We know this because Kendall's contribution to Mrs. S was through an ovum, which can only be formed from an XX genome — Mrs. S was not formed from Kendall's sibling's cells. Mrs. S also shares the same amount of DNA (50%) with the Leda clones, since they are clones of the XX genome. On the other hand, Mrs. S shares only 25% of her genes with the Castor clones. Because of this, they are biological uncles to Mrs. S — clones of her mother's brother.

Mrs. S is the child of the XX genome, as is Kira, Sarah's daughter, so on the genomic family tree they occupy a similar position. Kira shares 50% of her DNA with the Leda clones and Kendall's XX cells and 25% of her DNA with the Castor clones. Therefore, Kira and Mrs. S have the same genetic relationship as siblings, sharing 50% of their genes with each other. Sibling genetic similarities are more variable than those between parent and child, ranging from 25 to 75%. Nonetheless, it starts to make your head hurt when you realize that Kira and Mrs. S could be considered genetic siblings.

Finally, let's make at least one thing clear: it isn't really correct to start referring to Sarah as Siobhan's "mother." We've seen this mentioned more than once in online discussions. Calling Sarah Mrs. S's mom just because she was cloned from Kendall is kind of like saying

PROJECT CASTOR

Shares roughly
50% of DNA

PROJECT LEDA

Shares 100%
of XY DNA

Shares 100%
of XX DNA

KENDALL

Shares roughly
25%
of DNA

Shares
50%
of DNA

Shares
50%
of DNA

Shares 50% of DNA
most likely from XX DNA

MRS. S

KIRA / HELENA'S BABIES

Shares roughly
25% of DNA

ORPHAN BLACK
FAMILY TREE

that Alison, Cosima, or any of the other Leda clones is Kira's mother by virtue of being genetically identical to Sarah. That's just . . . not how it works. Besides, as *Orphan Black* shows us, genetics aren't the bottom line, and chosen family and familial relationships are just as, if not more, important than genetic ones. (Meaning Mrs. S is Sarah's mum, and that's final!)

RACHEL

We may have been raised without familial bonds, but you, Ira, are my brother. And I'm beginning to believe that means something.

(4.08 "The Redesign of Natural Objects")

Now, with Kendall's multitude of familial and genetic relations, one obvious question comes to mind: why don't the Leda clones look exactly like Kendall? After all, Kendall is a woman and the Leda clones are genetic identicals of Kendall's XX genome, so it is should follow that the clones and their original would look exactly alike. Right? Wrong, actually. It's Kendall's uniqueness as an original for two different lines of clones that explains why she doesn't look exactly like either set of clones. As a chimera, Kendall's phenotype (as in her physical traits and appearance) is determined by two separate genomes. A majority of her phenotype is determined by the XX genome (the Leda genome) because she is female and XX was her initial genome during development; however, with the absorption of her brother in utero and the incorporation of the XY genome, she now has the XY genome influencing her overall appearance. This is why she does not look like the Leda clones: her appearance is a combination of two genomes while the Leda clones all have only one genome. If Kendall weren't a chimera, and her brother had been born separately, she would look like the Leda clones, and her brother would look like the Castor clones. Cosima and Scott nerd out over this fact together at Alison's celebration dinner in "History Yet to Be Written," but they use genetics terms. The others don't understand just how fascinating it is that the Leda clones don't look like perfect copies of Kendall.

> **SCOTT**
> Yeah, because in a chimera, each
> cell line influences the phenotype
> of the whole. [...]

> **FELIX**
> Donnie, you're a civilian.
> Would you please say something
> understandable.

> **DONNIE**
> The frickin' beef is the bomb.

> **FELIX**
> Thank you.

(3.10 "History Yet to Be Written")

BUT WHAT ABOUT —

So many of the questions that we've received from Clone Club revolve around Kendall Malone — in particular the season four revelation of Kendall's leukemia and its implications for the Leda clones and Cosima's treatment.

In "Transgressive Border Crossing," after convincing a reluctant Kendall to donate her samples for research toward the clone disease cure, Cosima learns that Kendall is suffering from leukemia, a cancer of the blood-forming tissue of the bone marrow and lymph nodes. Most commonly in patients with leukemia, the white blood cells are the most affected and are therefore not able to properly fight off infections.

In Kendall's case, the most heartbreaking part of this news is that she is sick and could potentially die so soon after being reunited with her daughter. Of course, this point becomes moot when Detective Duko shoots her and then incinerates her body. However, Kendall doesn't just have herself to think about; how *does* her leukemia affect the Leda clones?

Well, the answer is, maybe not much at all. Cosima was using Kendall's cells in her research, but the fact that they could potentially be cancer cells wouldn't have made much of a difference. Cosima needed Kendall's biological material mainly in order to compare Kendall's genome with the clone genome to identify the infertility sequence. She would have been able to successfully do this with almost any cells from Kendall. Even if the cancer cells contained genomic anomalies, as is common with cancer cells, Cosima could have simply used samples from unaffected cells. Of course, at the very same time that Kendall was being killed, Sarah acted on impulse to destroy their only sample of Kendall's cells with bleach (the cells' digital information having been wiped from their computers by Neolution after their only other copy of the genome, along with Cosima's research notes, had been handed over to Susan Duncan).

BY A MATTER OF FATE AND UNUSUAL BIOLOGY, KENDALL BECAME THE MOST CRUCIAL PLAYER IN THE DEVELOPMENT OF THE LEDA AND CASTOR CLONES EVEN THOUGH SHE WASN'T EXACTLY A WILLING PARTICIPANT.

The main worry then becomes whether Kendall's leukemia is genetic. Cancer can be caused by environmental factors, such as diet, exercise, or smoking, and if that is the case with Kendall's leukemia, then Clone Club doesn't have to worry about it too much. Kendall is a cigarette smoker and has probably been a smoker for majority of her life. However, we know that the leukemia was only in the Leda cells in Kendall's body, which is how Scott was able to isolate the just-Leda cells from her samples. So, there is a chance her leukemia is genetic, putting all the Leda clones, and to a lesser extent Kira, at risk for developing the disease.

Kendall Malone is a reluctant hero in this story. By a matter of fate and unusual biology, she became the most crucial player in the development of the Leda and Castor clones even though she wasn't exactly a willing participant. The clones owe their lives and genetics largely to her. When we meet her in London, she continues to resist being a part of the drama. It takes a combination of hope — the possibility of rebuilding bridges with her daughter Siobhan (who is likewise resistant to this idea at first) — and gentle coaxing from Cosima to donate yet more tissues to benefit the clones, people born of her and yet strangers to her. It's a tragic fact that she is killed and her cell samples destroyed so quickly in season four, but the impact of her presence on the clones, physical and biological, remains.

CASE STUDY:
ELIZABETH CHILDS

ID: 317B31
DOB: April 1, 1984
Birthplace: Toronto, Ontario, Canada
Status: Deceased

FELIX

Sarah, who is Elizabeth Childs?

SARAH

I don't know. Just a girl who looks
like me. A girl with a pretty nice
life.

(1.01 "Natural Selection")

We don't know much about Beth Childs before she became self-aware as a clone; after all, the show begins with her death, and from there we only get to know Beth through flashbacks, fever dreams, and Sarah's attempts to impersonate her. One key thing we do know about the mysterious cop that led us into this adventure is that she had a problem with abusing prescription drugs. Psychopharmaceuticals — drugs taken to treat psychological disorders such as depression, anxiety, insomnia, et cetera — work by adjusting the activity of various chemicals in the brain, and abuse of these drugs can have very harmful effects on mood, personality, and the health of the patient.

The most common types of psychopharmaceuticals prescribed to patients with depression and anxiety disorders are the selective serotonin reuptake inhibitors (SSRIs), the serotonin-norepinephrine reuptake inhibitors (SNRIs), and the monoamine oxidase inhibitors (MAOIs). These drugs interact with various chemical pathways in the brain to affect the behavior of neurotransmitters.

Neurotransmitters relay signals from one neuron to the next, passing along messages through a neuronal circuit; the brain's response can affect mood, behavior, and so on. Serotonin and norepinephrine are two neurotransmitters that psychopharmaceuticals target. Reuptake inhibitors — SSRIs and SNRIs — prevent the reabsorption of neurotransmitters by neurons. Normally, after a chemical signal is transmitted from one cell to the next, the neurotransmitters are reabsorbed by the cells in order to prepare for the sending of another signal.

SSRIs and SNRIs cause the neurotransmitters to remain in the space between the sending and receiving neurons, known as the synapse. By maintaining these higher levels of the neurotransmitters in the synapse, the cells have an easier time communicating with each other, strengthening the neuronal circuits in the brain and thus strengthening the body's response (that is, improving mood, alleviating exhaustion and fatigue, and benefiting a patient's overall well-being). MAOIs, on the other hand, block the activity of the enzyme monoamine oxidase, which breaks down neurotransmitters, thus rendering them nonfunctional.

There's a similar result as with reuptake inhibitors: increased levels of neurotransmitters and thus stronger neuronal circuits.

Abuse of these drugs can be incredibly dangerous and have long-term effects. While on the drug, neurotransmitter levels are high and neuron communications are strong. When the drug leaves the person's system, these levels drop and neuronal functions are weakened. Over time, circuits can begin to respond less to the drug and have bigger crashes once the drug is out of the system. When the drug is taken at the proper dosage, this should not happen. However, abuse of the drug creates a state similar to what occurs with recreational drug use — a need for higher levels of the drug to receive the same effects, or high and drastic crashes when there is no drug.

Psychopharmaceuticals require careful management. A missed dose can result in drastic mood swings, headaches or migraines, or nausea. Similar effects occur if someone suddenly stops a dosage without properly weaning off. The symptoms mirror those of recreational drug withdrawal: fever, vomiting, muscle spasms, migraines, et cetera. These symptoms can trouble abusers when they attempt to stop or adjust their overuse, making it harder for them to get out of the cycle.

Although the general recommended doses of these drugs are aimed at minimizing adverse side effects, cyclical highs and lows, and dependency, psychopharmacology is very personalized. Not every drug works the same from person to person, and some people will have fewer side effects on one drug versus another. Occasionally, patients on certain drugs or at certain dosages even experience effects opposite to those intended. Brain chemistry is very complicated, and it can take time to find the right combination and dosage of drugs to treat an individual.

These drugs are highly complicated, and it's extremely difficult for people like Beth to manage them; we see her begin to abuse and eventually become addicted to psychopharmaceuticals. Beth's struggles with pills and her mental health were not, however, what led her to that train platform the night we first meet her. Evie Cho was actually the one to convince Beth that suicide was the only way to be free from Neolution. It's a good lesson on how mental health — and the use of drugs to assist with mental health — is complicated, and on how assumptions shouldn't be made about people dealing with these issues. Beth Childs was complicated: her life, her mental state, her job, and everything she dealt with were massive sources of stress. Given all these factors, coupled with her drug abuse, it's a wonder she was ever able to navigate and unravel the great mysteries of *Orphan Black*.

CASE STUDY:
TONY SAWICKI

ID: 320C65
DOB: March 31, 1984
Birthplace: Cleveland, Ohio
Status: Alive

TONY

Look at us. We're hot. Damn, girl.

SARAH

Not our usual identity crisis.

TONY

I did all that work a long time ago.

(2.08 "Variable and Full of Perturbation")

"Variable and Full of Perturbation" introduces us to Tony Sawicki. After his friend (who we later learn was his monitor) dies from a gunshot wound, Tony travels to Toronto in search of Beth Childs. He eventually comes face-to-face with Sarah, learning that he is a clone and sharing a bit of his carefree self-confidence with her.

Tony is overtly flirtatious, rocks a wild mullet, likes testing people's boundaries, enjoys a beer (or six), and is the only known male Leda clone. Tony is a trans man, meaning he was assigned female at birth but is a man. As part of his transition from female to male, Tony takes shots of testosterone. As he administers a shot into his leg in Felix's apartment, he explains how he never misses a dose, no matter what the situation or where he finds himself.

Testosterone is a hormone produced in the testes and, to a lesser extent, in the ovaries as well as the adrenal gland. During embryonic development, testosterone promotes the formation of the testes and penis; in adults, it's responsible for the formation of sperm.

Testosterone also promotes what are considered to be male characteristics: deep voice, facial hair, high levels of muscle formation, and
a high libido.

TONY

Hey, you sure you don't want a hit?
It'll put some hair on your chest.

(2.08 "Variable and Full of Perturbation")

Androgen replacement therapy (ART) is the use of exogenous
testosterone to replace estrogen and make a person's hormone
levels represent those of a "typical male." ART changes many of
the external characteristics and secondary sex characteristics of a
person, causing the growth of facial hair, a deepening of the voice,
increased ability to build muscle, a redistribution of body fat, changes
in body odor, increased body hair, oily skin and acne, cessation of
menstruation, increased libido, growth of the clitoris, and loss of hair
on the head. These changes all occur at different times and rates on
an individual basis, and to different degrees, but testosterone has the
ability to change a person's body in all of these aspects. For many
trans people taking testosterone, these changes aided by ART are a

necessary part of their transition that helps their appearance more closely match their gender and gender expression.

In the case of Tony, his use of testosterone has allowed him to grow facial hair and deepened his voice. ART won't cause breasts to go away (although breasts can become less firm as fat is redistributed), so Tony wears a binder to flatten his chest. ART will cause menstruation to stop after a few weeks to a few months, and menstruation will not begin again unless the ART is ended. However, testosterone will not affect the genitalia beyond a certain degree; the clitoris will become enlarged, but it will not resemble a penis. Some trans men, Tony included, choose to wear a packer to add a bulge, and in some cases the packer is functional for urination and sex as well.

There are some side effects associated with ART that are negative and potentially life-threatening. Taking testosterone can increase blood pressure, which can increase the risk for heart attack and stroke. Testosterone can also increase the incidence and severity of headaches and migraines. A primary danger of testosterone therapy is that it may increase the risk of some cancers. Although the connection isn't definitive, there is some evidence that there is an increased risk of estrogen-sensitive cancers, such as breast, ovarian, and uterine cancers, in people who take testosterone, as some of this testosterone gets converted to estrogen, which can then lead to abnormal cell growth in the sensitive tissues. It's often recommended that trans men undergo a prophylactic hysterectomy and oophorectomy (removal of the uterus and ovaries) by the time they have been on testosterone for five years to reduce any potential risk for cancer in these organs. However, no long-term studies have been performed due to a lack of funding and patients.

As for Tony, he appears to be healthy and to experience no negative effects from his testosterone. Although we don't know how long he has been on ART, surgeries don't appear to be on the forefront of his mind. After all, witnessing your best friend get shot, going on the run, finding out you're a clone, and then attempting to disappear from the watch of a global sinister corporation would take up the majority of your time.

CHAPTER 4

"THIS IS MY BIOLOGY, MY DECISION"

SYNTHETIC BIOLOGY AND HUMAN EXPERIMENTATION

LEEKIE

Neolution gives us the opportunity
at a self-directed evolution and I
believe that's not only a choice,
but a human right.

(1.06 "Variations Under Domestication")

We get our first glimpse of Neolution through Cosima's eyes as she sits in on a lecture given by Dr. Aldous Leekie in season one's "Variations Under Domestication." It's an interesting choice for an unofficial first date with (apparent) fellow grad student Delphine Cormier, followed up with post-lecture wine theft: How to Win Cosima's Heart 101. Cosima is incredulous at many of Neolution's claims and isn't shy at all about spouting cynical retorts during Dr. Leekie's lecture. But at the same time, she can't help but be drawn in by the ideas and philosophies that the movement puts forth — to the point that Sarah claims that Cosima "drank the purple Kool-Aid." But it's easy to see the aspects of Neolution that appeal to Cosima the scientist (she's studying evolutionary development, or evo-devo, after all). Neolution seeks to put evolution in the hands of the individual, to improve upon natural human design, and to grant enhanced or even superhuman capabilities. Who wouldn't want that? Of course, in season one, neither the audience nor Cosima knows to what extremes Neolution has gone, and is willing to go, to achieve these ends.

NEOLUTION

Neolution has its roots in eugenics and transhumanism. Sometimes also abbreviated as H+, transhumanism is a movement dedicated to developing and implementing technologies to enhance and improve human beings and to transcending humans' intellectual, psychological, and physical limitations. If this sounds to you like something from a science-fiction novel, you're not wrong: transhumanism isn't a new idea. Aldous Huxley, whose dystopian novel *Brave New World* we've already mentioned, was an early advocate for transhumanism. (Dr. Leekie's first name is absolutely a nod to this famous Aldous while his last name is most likely a reference to the paleoanthropologist family the Leakeys — specifically Louis Leakey, who believed that paleoanthropology was necessary to unlocking the secrets of human evolution.) However, it is Huxley's pal British geneticist J.B.S. Haldane who is credited with establishing the fundamentals of transhumanism in the 1930s. Haldane's main interests included eugenics and using genetics to improve human health and intelligence, as well as ectogenesis (gestating fetuses in artificial wombs). If Haldane and Leekie had lived at the same time, they probably would have been best friends and lab partners.

In season four's "The Antisocialism of Sex," Rachel — held hostage in a room on the Neolution Island for most of the season — suddenly and unexpectedly finds her door left open. She struggles up a steep set of stairs that leads her to a leather-bound book titled *On the Science of Neolution* that features images of cells dividing, a portrait of Charles Darwin, and a painting of Leda and the Swan. Susan finds Rachel poring over the book and informs her that Neolution was founded by industrialist Percival T. Westmorland in the mid-1800s, around the time Charles Darwin published his theories of evolution

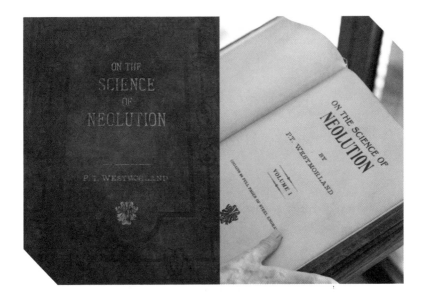

in his famous work *On the Origin of Species* (the full original title of which is *On the Origin of Species by Means of Natural Selection, or the Preservation of Favoured Races in the Struggle for Life,* which certainly has more of a Neolutionist ring to it). In the season four finale, we discover that Westmorland is apparently still alive in 2013.

Westmorland is not a real historical figure, but that isn't to say that people who shared in his ideas and ideals for human progress didn't exist in the mid-1800s. In an era of rapid progress in science and medicine, people were beginning to wonder about the implications for humans and their lives. Mary Shelley's *Frankenstein* (subtitled *The Modern Prometheus*) was published in 1818 and drew inspiration from then-current science as well as another Darwin: according to Shelley, Dr. Erasmus Darwin, Charles Darwin's grandfather, animated a piece of vermicelli in a glass plate. This is a funny misunderstanding on Shelley's part: Darwin never observed a noodle gain mysterious new life. Rather, Darwin had spent time observing *vorticella*, waterborne protozoa, and found that, although it would appear inactive during months kept in a dry state, it appeared to survive and revive once returned to water. In general, Erasmus Darwin believed in spontaneous generation — the notion that living organisms could develop spontaneously out of inanimate matter like dust or dead flesh, and that microbes (then called animalcules), specifically, could be born of dead matter. Other scientists of this era who inspired *Frankenstein* include Luigi Galvani, who experimented with "animal electricity" created in the brain and sent to animate muscles and organisms via nerve pathways, and Humphry Davy, a chemist and philosopher (and friend of Shelley's father) who had big ideas about the progress of science. As Davy wrote in *Elements of Chemical Philosophy*, "science has [. . .] bestowed upon man powers which may be called creative; which have enabled him to change and modify the beings around him."

Transhumanism today seeks to transcend the limitations of human biology. Think about how technology has evolved over your lifetime — the computers and phones that you used ten years ago, even two years ago, for example — and how much has changed and improved. Comparatively speaking, biological evolution is much slower and less efficient than technological evolution. It follows, then, that technology can be employed to enhance human biology and push it to the next level. This idea is at the root of Neolution, the driving force behind the new technologies that we see on *Orphan Black*. Everything from Cosima's glasses to Rachel's cyborg eye is a technology to surpass human limitations. While scientists worked to perfect manipulations of the genome, Neolutionists took matters into their own hands and began changing their bodies to reflect evolutionary traits that they would like to possess, should they find a way to manipulate their own genomes. One white eye, tails, bifurcated genitalia — followers of Neolution take on any modifications they see fit in the name of "directed evolution," or, as Aldous Leekie calls it, "self-directed evolution."

> FOLLOWERS OF NEOLUTION TAKE ON ANY MODIFICATIONS THEY SEE FIT IN THE NAME OF "DIRECTED EVOLUTION," OR, AS ALDOUS LEEKIE CALLS IT, "SELF-DIRECTED EVOLUTION."

One branch of transhumanism includes biohackers — basically the sort of crowd you might run into at Club Neolution, "Freaky Leekies" who apply the philosophies of transhumanism and

Neolution on a smaller, individual scale. Many real biohackers take a DIY approach to self-enhancement. They might experiment with diet or sleeping patterns in order to enhance mental ability, or they might undergo minor surgeries, such as implanting magnets under their skin. In "The Collapse of Nature," Beth meets a young woman named Trina at Club Neolution who has implanted magnets into her fingertips and demonstrates their use by magnetically lifting a pen from Beth's shirt pocket. The usual intent behind implanting magnets beneath the skin is to be able to sense magnetic fields, thus gaining a sense that humans usually lack.

> **LEEKIE**
> Neolution is an open-source concept.
> We embrace the fringes, the fiction
> to our science.

> **BETH**
> Your book inspires them to play God,
> to tinker on themselves in basement
> labs.

LEEKIE

Every scientific discovery in
history, including God, began with
tinkering in the basement.

BETH

God was a scientific discovery?

LEEKIE

Little joke. We're all impatient
for the future.

(4.01 "The Collapse of Nature")

Most of the Neolutionists that we meet in *Orphan Black* take the concept of transhumanism to the next level by using advanced science and technologies to enhance human abilities in a permanent and potentially heritable way. On the simpler end of this scale, we have Olivier, who opted to undergo surgery to gain a functional tail (that Helena ultimately cuts off in season one). On the more intense and involved end, we have Evie Cho's cyborg maggots, meant to infect their hosts and deliver transgenes that alter host DNA; we have Brightborn babies, produced by germline editing (more on that below); and we have Susan and Ethan Duncan's Projects Leda and Castor. All of these are very different approaches to reaching the next step in human evolution, and each comes with its own consequences.

Directed evolution isn't an idea exclusive to *Orphan Black*. In the realm of protein engineering, scientists use directed evolution to mimic natural selection in order to evolve proteins toward a specific goal. This allows for genetic diversity in an experimental setup. In the case of Neolutionists, they are not affecting their genetics, but they are changing physical traits to achieve qualities that would take hundreds of thousands of years to attain via natural selection. In this way, they are directing their evolution by bypassing genetics to gain the appearances and enhancements they want.

Although transhumanism and Neolution have cast a shadow over the plot of *Orphan Black* since season one, there's a special emphasis on these ideas in the show's fourth season. The episode titles for season four were inspired by the works of Donna Haraway, a science philosopher and essayist whose works — perhaps most famously "A Cyborg Manifesto: Science, Technology, and Socialist-Feminism in the Late Twentieth Century" from her book *Simians, Cyborgs and Women: The Reinvention of Nature* — have clear ties to the show's themes, especially in terms of casting a critical eye toward topics in body and identity politics and feminism. Cosima Herter, *Orphan Black*'s science consultant, recommended "A Cyborg Manifesto" to show cocreator Graeme Manson while he was developing the series and looking for historical and philosophical inspiration. In a blog post for *Orphan Black* titled "Eat Me/Drink Me and the Trouble with Cyborgs," Cosima (the real Cosima) writes:

> We've used [science] to problematize questions of identity, individuality, autonomy, agency, feminism, and freedom. This season, we continue to problematize these ideas but we've veered a bit into kindred disciplines — technology and engineering. We've opened some very uncomfortable arguments about the industrialization and commodification of biology through — and as — biotechnology. We've ripped open some assumptions about bioengineering, biomedicine, reproductive technologies, and how biology-for-profit might affect questions of kinship, autonomy, propriety, identity, and what we consider the basic building blocks of life.

Haraway defines *cyborg* as anything that transgresses boundaries, whether by technologies or other means, and she uses her work to explore nature and culture, what defines human versus nonhuman versus almost human, and different notions of property (who owns a body, anyway?). Many of the episode titles in season four have roots in Haraway's essay "The Biological Enterprise: Sex, Mind, and Profit from Human Engineering to Sociobiology," which looks specifically at human engineering and capitalism — themes explored through Evie Cho, who acts independently of Neolution to commodify genetically modified humans and to secretly launch her own eugenic projects through Brightborn's offerings. Part of Haraway's essay focuses on American psychologist, primatologist, and eugenicist Robert Yerkes, who envisioned human engineering to benefit culture, society, and industry, and who described it in a way that almost sounds like an overview of Project Leda:

> *It has always been a feature of our plan for the use of the chimpanzee as an experimental animal to shape it intelligently to specification instead of trying to preserve its natural characteristics. We have believed it important to convert the animal into as nearly ideal a subject for biological research as is practicable. And with this intent has been associated the hope that eventual success might serve as an effective demonstration of re-creating man himself in the image of a generally acceptable ideal.*
>
> (*Robert Yerkes,* Chimpanzees: A Laboratory Colony)

Ultimately, Yerkes is best known for his work developing intelligence tests for World War I conscripts, and for using those same tests for various racist purposes before and after the war.

Haraway argues that human engineering and commodification alter how humans think about and experience relationships; for example, suggesting a genetic component to illness or undesirable traits allows us to perceive people who have these illnesses or traits as having a fault in their code and to believe that, if we could just pinpoint the error, we could fix it and correct the code. While the intent might be good (improving a person, or a person's DNA), it is inherently flawed in that it ignores the person as an individual in favor of a universal ideal. The person becomes a product, a mere container for genes.

BRIGHTBORN TECHNOLOGIES: MAKING A BETTER TOMORROW TODAY

The main entity antagonizing the clones in season four is the Neolutionists, once again following the science toward progress and

genetic manipulations. One Neolution branch comes in the form of Brightborn Technologies, which masquerades as a fertility clinic but is really a secret program for rearing babies with genetic abnormalities — a combination of poor ethics and illegal science with the goal of, as far as we can tell, improving the human race. While their motives are questionable and their methods evil, the scientists behind Brightborn and Neolution follow a strict experimental outline on the road to a new future.

Although Brightborn does uphold their facade of a fertility clinic by helping infertile couples to have babies, the real purpose of the clinic is to explore and understand genetic diseases and abnormalities. To do this, the scientists manipulate specific genes of an embryo, then hire women to carry the fetuses to term. In the season four episode "Human Raw Material," Cosima manages to sneak into Brightborn in hopes of uncovering the mysteries behind the clinic, and she witnesses a horror she could not have imagined: some of the children of Brightborn carriers suffer from rare and severe genetic disorders. Cosima dons scrubs and steps in to observe the birth of a Brightborn baby; the baby is born with popliteal pterygium syndrome.

Popliteal pterygium syndrome, or PPS, is a rare genetic condition that affects the development of the face, skin, and genitals. It is most frequently characterized by a cleft lip, cleft palate, or both. Sometimes this clefting is so severe that a depression or pit forms in place of certain facial structures. The baby that Cosima sees has a severe form of PPS: she is floppy-limbed, and her nose and mouth are unformed and appear as a large hole, which causes her to experience difficulties breathing; the baby is quickly terminated. By giving these children rare diseases, the scientists at Brightborn are able to understand the effects of manipulating a given gene, thereby figuring out which genetic changes can be used to improve the human condition and which ones have deleterious effects. The ideal changes are then applied in the clinic, giving the paying parents the "best" child science can make.

"WE'RE YOUR ONCOMICE": ESTABLISHING A BASELINE

When it comes to something as daunting and complicated as manipulating the human genome, it is important to start with a point of reference — something with which to compare all experiments and data. With such diversity in the human population, how do scientists decide which people should be used as a standard control? Enter Project Leda, a massive project involving near-perfect replicas of a human genome, a seemingly infinite source of material to establish reference points with which to measure all future manipulations.

As Evie Cho says, the clones provided a baseline for the scientists, the reference points they needed before moving on to bigger and better genetic challenges. With the addition of known synthetic sequences, the clones allowed the scientists to see how a natural human genome would hold up with manipulations over time. Interestingly, if the Leda and Castor clones are their only evidence, then the answer to this line of inquiry is: not so well at all. It isn't until the clones are receiving treatments to correct the disease-causing mutation that Coady remarks that Project Leda can finally perform as an ideal baseline again.

What exactly are synthetic sequences? As the term implies, these are strings of DNA that are not natural parts of the genome. The Dyad scientists created these sequences in the lab and then inserted them into Kendall's genome to create a novel clone genome. Some of these sequences, such as the ID tag sequence used to identify each clone, are noncoding (they do not represent a gene and therefore are not functional sequences). Other synthetic sequences, however, such as the infertility sequence, encode functional genes that are transcribed and therefore have an actual effect on the clones. The infertility sequence causes the ova of the Leda clones to never fully mature, preventing them from getting pregnant. As a side effect, the clones also develop the clone disease.

While there are currently no known cases of humans with synthetic sequences, synthetic DNA isn't science fiction. Scientists create synthetic DNA sequences in the laboratory to study specific genes, as a means of studying genetics without the need for a constant source of biological material. Synthetic genetics has even grown to the level of synthetic microorganisms — entire single-celled organisms made completely in the lab — which has allowed for large-scale development of organisms as biofuel, microorganisms to combat oil spills, and the creation of platforms to study immunotherapy and cell therapy.

Considering that the clones were born in the mid-1980s, the scientists behind their creation were most likely unsure of how synthetic sequences would hold up in the human genome over time, and whether or not they would even be properly integrated and expressed. Therefore, the Leda clones were a necessary first step before the scientists moved on to modes of integrating synthetic sequences into people at will; Project Leda was an experiment to test the viability of their methods. Unfortunately, and unethically, this involved the lives of dozens upon dozens of clones who didn't consent to the experiment.

In "Human Raw Material," Cosima compares the Leda clones to OncoMouse, a creature who, by genetic design, was meant as a research tool. OncoMouse sparked controversy in the 1980s and '90s over what it means to patent living organisms (a familiar theme on *Orphan Black*) and the ethics of gene editing and germline editing. The first animal ever to receive patent protection, the OncoMouse is a type of transgenic laboratory mouse that has been genetically modified to carry a human oncogene (genetic material with the potential to cause cancer), and, as such, it is highly susceptible to developing cancerous tumors. This susceptibility to cancer is also heritable, which makes these mice useful as models for cancer research. As Susan Duncan's "OncoMice," the Leda clones are a genetic template for researching the human genome and identifying strategies for engineering healthier, superior humans. They are also, like the OncoMouse, patented property.

The debates and controversies over patenting organisms may have begun with OncoMouse in the 1980s, but the conversation has endured longer than OncoMouse's patent (it expired in 2005, and, honestly, if the Leda clones were patented in the 1980s, their patent has probably expired as well). In 2002, a treaty initiative to share the genetic commons was launched in an effort to control the commodification of genes and organisms in both natural and synthetic forms and to prevent monopolies created by gene patents. This treaty was one of the first to suggest global protection of clones:

> *Therefore, the nations of the world declare the Earth's gene pool, in all of its biological forms and manifestations, to be a global commons, to be protected and nurtured by all peoples and further declare that genes and the products they code for, in their natural, purified, or synthesized form as well as chromosomes, cells, tissue, organs, and organisms, including clones, transgenic, and chimeric organisms, will not be allowed to be claimed as commercially negotiable genetic information or intellectual property by governments, commercial enterprises, other institutions or individuals.*
>
> *(Treaty to Share the Genetic Commons, 2002)*

In the season two episode "Nature Under Constraint and Vexed," Rachel Duncan has a phone conversation (in German) with a stakeholder about how "most of the patents were submitted immediately after the decision of the Supreme Court." Later in the same episode, she tells a group of Korean stakeholders that "the recent Supreme Court decision characterizing the legal status of natural versus synthetic DNA was the successful result of our lobbying strategies. We are proceeding with the next tranche of patent claims." She's referring to the June 2013 U.S. Supreme Court case of *Association for Molecular Pathology v. Myriad Genetics, Inc.*, in which the court ruled that human genes could not be patented in the United States because DNA is a product of nature and not an invention. (As a fun

timeline-related aside, this episode aired in April 2014, and so Rachel was commenting on relatively current real-world events. However, in season four, Felix makes an offhand comment about only six months having elapsed since Sarah saw Beth die at Huxley Station, which places season four's events in around May or June 2013 — the pilot takes place in late November 2012 — before the ruling on gene patents occurred. *Orphan Black* operates on an alternate timeline to that of our universe, so it seems that the Supreme Court in *Orphan Black* just happened to make their ruling regarding gene patents a few years ahead of us.)

Years earlier, Myriad Genetics had been granted patents on the BRCA1 and BRCA2 genes, whose expression has been identified as being strongly linked to aggressive forms of breast and ovarian cancers. One of the major issues of granting a patent on a gene is that it places control over research involving that gene in the hands of the entity holding the patent. Any person wanting to conduct research — say to do studies that might lead to more effective testing or treatments for BRCA-linked cancers — would have had to have paid steep royalty fees to Myriad Genetics. As you can see, gene patents can restrict important progress, although it could also be argued that without such patents, Myriad Genetics might not have been able to afford to craft tests for BRCA1 and BRCA2 mutations in the first place.

In season one, we learn that the Leda clones are considered patented products in the *Orphan Black* universe. Cosima and Delphine make this discovery in the season one finale, "Endless Forms Most Beautiful," when they decode a barcode sequence in Cosima's genome, which reads "This organism and derivative genetic material is restricted intellectual property."

COSIMA

That synthetic sequence, the barcode I told you about? It's a patent. [...] We're property. Our bodies, our biology, everything we are, everything we become, belongs to them. Sarah, they could claim Kira. They patented us.

(1.10 "Endless Forms Most Beautiful")

But, in reality, could the Leda clones be considered patented intellectual property? Well, first, and most importantly: patents are public. So, unless the people who filed the Project Leda patent were exceedingly clever and managed to mask the fact that the patent was for *human clones*, this patent would have made the clones' existence public knowledge. Clearly, that isn't the case. But, for the sake of argument, let's ignore that fact.

SCOTT

Back in the early '80s, genetics was like the Wild West. No one knew the science. The laws didn't exist, so they slapped provisional patents on everything.

COSIMA

Like pissing on a fencepost.

(2.05 "Ipsa Scientia Potestas Est")

In Canada, one of Dyad's base locations, patents have been allowed on genes and individual cells since the 1980s, but not on entire organisms — like genetically modified mice — and definitely not on cloned humans (plants are A-OK to patent, though!). Although the U.S. did grant a patent for OncoMouse in 1988, Canada refused the patent. In 2002, the Canadian Supreme Court's interpretation of the word "invention" officially decided that higher life forms would be unpatentable. Currently, the Canadian Assisted Human Reproduction Act also bans the creation of human clones by any technique, but this only came into effect in 2004 — not early enough to have affected the original Projects Leda and Castor, but at least one year before Charlotte was born.

Despite its ruling genes unpatentable, Rachel states that the U.S. Supreme Court decision was ultimately a success for Dyad. This may be because the ruling only covers naturally occurring genes; synthetic sequences, like those included in the Leda clones' genome, may still be patented, and any successful human cloning techniques that the Project Leda team invented could be patented as well. But the clones themselves and any children that they might have? Although there isn't a precedent for it, in the U.S. the clones' rights as human organisms would probably supersede their patents as intellectual property. Rachel acknowledges this in "Clutch of Greed" when she tries to convince Sarah to allow Dyad to study Kira's unique biology.

SARAH

She's a little girl. Why? What are you looking for?

RACHEL

Anything we find will of course be proprietary. But she isn't. She's yours.

(5.02 *"Clutch of Greed"*)

This is why it is so important to Rachel in the season four finale that they relaunch human cloning in countries not only where it is legal (or, rather, not *illegal*), but also where the next generation of clones' status as intellectual property would supersede their status and rights as people.

THE MAGGOT-BOT

The maggot-bot is another one of Evie Cho's genetic (and probably eugenic) projects. In the season three finale, "History Yet to Be Written," Dr. Nealon reveals to Delphine Cormier that Neolution, the "self-directed evolution" movement that Aldous Leekie represented in season one, is the controlling force behind Projects Leda and Castor. He warns her that Neolution science is far more advanced that she would ever guess. Then something even more horrifying happens: Nealon attacks Delphine and opens his mouth to reveal a bloody, maggot-like creature — it looks like he's attempting to infect her with it. Delphine manages to kill him before that can happen, and she pockets the maggot as a sample. No explanation for the maggot's existence or purpose is given.

That is, until the season four premiere, "The Collapse of Nature," when we are whisked back in time to when Beth was alive. The season opens with mysterious hacker clone M.K. discovering two "cheek choppers" burying a body in the woods. The body has the flesh of its right cheek removed; it also has one gray contact and a bifurcated penis (split in two down the shaft). M.K. relays this intel to Beth, who has been investigating Neolution and how they connect to the clones. Beth's search leads her to Club Neolution and Trina, a young Freaky Leekie whose boyfriend has received a strange implant in his cheek. According to Trina, the implant is meant to be a biometrics tool, recording data from his body.

The thing is, the implant is *growing* — it's the very same type of modified maggot that we saw escape Nealon's mouth the previous season.

And then, in "Transgressive Border Crossing," Sarah meets a new source, a hacker called Dizzy, who mistakes her for M.K. and shows her a nightmare-worthy video of the Neolution implant in action. Dizzy describes it as "organic, gene-spliced, maybe." While it's unclear at that point exactly what the implant is or what it does,

CLONE CLUB Q&A

I am interested in Nealon's Neolution implant. If it was in his cheek, how did he get it out? How was it meant to attack Delphine? I don't know, that was weird.

It *was* weird! Nealon's maggot-bot (seen only once in all of its strange and horrifying glory, in the season three finale) was our introduction to the implant, and its behavior seems to run counter to what is established later. The bots are extremely delicate and difficult to remove without triggering the release of a deadly neurotoxin, and yet we see Nealon's bot, bloody but intact, crawling from his mouth as if he intends to spit it at Delphine and infect her with it. What gives?

The maggot-bots have been around for longer than any of us might realize: Leekie already has his by the time we first meet him in season one (he jokes about having an implant, a neuro-lingual chip, that allows for him to speak and understand any language in "Variations Under Domestication," but meanwhile, he has a different implant entirely). In season three, it's clear that the implant has gone through many iterations and improvements over time: Evie received the newest version of the bot to treat her shingles, and Alison was almost infected with an older, apparently glitchy version of the bot when she was attacked in her home by the cheek choppers. It's possible that Nealon's version of the implant simply doesn't possess the neurotoxin kill switch and thus could be removed much more easily. Delphine did manage to grab the implant before escaping from Nealon, and perhaps she planned to study it (Delphine is an expert on parasite-host relationships after all!). Unfortunately (or maybe fortunately), we never do see that particular maggot-bot again. Maybe it was left behind in the parking lot where Delphine was shot, maybe it was confiscated when she was brought to Revival, or maybe it's somewhere at Dyad, waiting for a new host.

As for how Nealon's maggot-bot was meant to infect Delphine . . . well, Nealon's method of basically spitting it at her face is ineffective any way you slice it. Implanting the bot is far easier than removing it — as Cosima says, you basically cut a hole in your face and stick it in — and if the maggot has functional mouth parts, then cutting a hole first might not even be necessary. But unless Delphine had been unconscious when this all went down (and she wasn't), it would be fairly easy for her to avoid infection. It seems like Nealon's attack was more of a threat tactic than a real attempt to infect her, especially given that he warns her that someone else is on his way to kill her.

it's *definitely* alive and squirming inside its host's cheek; it appears to interact somehow with the host's blood. M.K. describes the implant as a "maggot-bot," which suggests that it has programmed functions. Moments after M.K. reveals this information, Sarah is attacked by the cheek choppers and realizes that she must have a living implant in her cheek, presumably placed there while she was being held by Dyad. Mrs. S later uses a flashlight to examine Sarah's cheek, and there it is: a maggot-bot attached to her blood vessels. And thanks to Dizzy's video footage, Sarah knows that when the worm is threatened, it unleashes some sort of defense mechanism that appears to kill its host.

In "The Stigmata of Progress," Cosima and Scott waste no time performing an ultrasound of Sarah's cheek to reveal that the implant is, in fact, an organic-cybernetic hybrid (a maggot-bot!), and that it's latched onto Sarah's maxillary artery. Dizzy suggests that this location was chosen for its proximity to the brain, which would seem to be the only good reason for placing it there (the cheek isn't exactly a well-protected place for an implant, and its host could easily bite into it, damage it, and trigger the defense mechanism). The maxillary artery

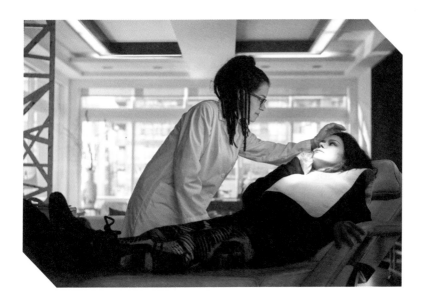

is a major artery, and its branches not only feed the face, jaw, and neck, but also enter the skull and supply the meninges around the brain.

The Neolution maggot-bot isn't strictly natural: it is a cyborg, a programmable piece of technology housed in a living organism, and it uses that organism's natural behaviors to inform its function — another example of Neolution combining advanced technology with natural processes. The maggot-bot seems to behave like a parasite in a way; parasites by definition benefit from living in a host, but the host shouldn't gain any benefits from being infected by the parasite.

Parasites don't usually kill their hosts, because they rely on the host staying alive to survive. Illness or death caused by parasites is usually thanks to the host's immune system trying to fight off the parasitic infection. The implant clearly works differently.

In nature, there is a distinct group of organisms known as parasitoids. Often parasitoids are lumped together with parasites, but the distinction between the two groups is that parasitoids ultimately sterilize or kill their hosts. There's a slew of ways: they can damage critical tissues and lead to organ failure; they can release toxins that can cause septic shock or enzymes that break into tissues and cause inflammation; they can relentlessly attack the host's immune system and exhaust it to the point that other microbes can move in and infect the host; or they can basically self-destruct in a way that also harms the host. Are there parasitoid maggots in nature? Definitely. Healthy caterpillars can be infected by parasitoid wasp eggs. The eggs hatch into larvae (maggots) that feed on the caterpillar from within, pupate, and emerge from the caterpillar's husk in adult form. The maggot-bot technology may be inspired by parasitoids in terms of their ability to kill their hosts, but, as far as we can tell, they don't seem capable of developing past a larval stage.

Dizzy's sources also lead Sarah to the office of a suspicious dentist specializing in implants, and it turns out that Beth had also uncovered this location. Hoping to get the implant safely removed (as one would and should!), Sarah instead winds up trapped by a Neolutionist dental assistant who threatens to puncture the maggot in her cheek with a sharp probe.

Turns out that if damaged, the implant releases tetrodotoxin into the artery. A powerful neurotoxin — you may know it as the poison found in pufferfish (and the reason why people have died eating improperly cut *fugu*, a prepared pufferfish delicacy) — tetrodotoxin works by disrupting the nerves' ability to transmit signals. Muscles become paralyzed, and death is usually caused by paralysis of the diaphragm and intercostal muscles: the muscles that you need to breathe. If you've been poisoned thanks to eating *fugu*, this can take anywhere from 20 minutes to a full day. The maggot-bot implant spills the toxin directly into the bloodstream, though, so it would kill you way faster, as we see in Dizzy's video.

Oh, and there's no antidote to tetrodotoxin.

Why would Neolution engineer an implant with this sort of kill switch? Most likely, it has nothing to do with the host or with the bot's main function. The release of tetrodotoxin is triggered when the implant is damaged, which is likely to happen if someone is trying to remove it from a living host. Neolution is likely protecting the design of the implant and limiting how much information can be obtained about it by killing the host. Because, as horrific as the Neolution maggot-bot might seem, its design and intended use are very helpful. We learn from Susan Duncan in "Human Raw Material" that the maggot-bot is actually a vector for gene therapy.

Currently, gene therapy is considered an experimental

treatment option, used only in severe illnesses such as cancers. For Neolutionists, however, gene therapy offers an avenue to overcome genetics and improve on the human condition. Evie Cho uses gene therapy to overcome the side effects of her autoimmune disease, a luxury not currently available for those suffering from severe combined immunodeficiency (SCID). Also known as "Bubble Boy disease," SCID is an immune deficiency that causes defects in T cells — that fight off active infections — and B cells — that produce antibodies and establish the adapted immune system. Because both branches of the immune system are compromised in SCID, people with this disease are susceptible to serious infections. Gene therapy has been used to treat SCID with varying degrees of success. While some trials have resulted in disease improvements and no adverse effects, other trials have resulted in multiple patients developing leukemia when the integration of the gene therapy either disrupts a tumor suppressor gene or activates an oncogene. In reality, gene therapy is still very experimental and uncertain, but in the world of *Orphan Black*, Neolution scientists used it to help Evie overcome her SCID and all subsequent side effects.

IN REALITY, GENE THERAPY IS STILL VERY EXPERIMENTAL AND UNCERTAIN.

Dr. Leekie, we also learn, used Neolution gene therapy to protect himself from his inherited genetic risk factors for Alzheimer's disease. Leekie's genetics, while not guaranteeing that he would get Alzheimer's disease, increased his chances of developing the debilitating and as yet incurable illness. His gene therapy inhibited the expression of the risk alleles, decreasing his chances of developing the disease.

The application of the Neolution gene therapy, however, was sinister in that these devices were tested in Neolution followers as they were developed, leading to outcomes that were often deadly. Most often the vector for gene therapy is an altered virus, but it could very well be a parasitic maggot or worm.

Remember the parasitoid wasp larvae? They have been of particular interest to scientists because the wasps have a mutualistic relationship with a family of insect viruses called Polydnaviridae. The virus's full genome is integrated in the parasitoid wasp's DNA, and it replicates in the calyx of the female and pupate wasps' ovaries. This replication creates particles containing segments of DNA. When a female wasp infects a host with its eggs, it also injects these particles, which enter into the host tissues. The viral genes packaged inside the particles are then expressed by the host cells. The host immune system then basically shuts down so that the wasp larvae can develop safely inside a living host without being attacked by a hostile immune system.

Scientists are looking to polydnavirus to create better gene therapy vectors. The particles can deliver 560 kilobase pairs of DNA (or kb, a unit of measurement in molecular biology equal to 1,000 base pairs of DNA), which is way more than what current vectors can carry. For perspective, of the more popular existing viral vectors, adenovirus can accommodate up to 38 kb, retroviruses can deliver about 8 kb, and herpes simplex virus can deliver up to 150 kb.

At least one other parasite has been looked at closely and considered a potential tool for gene therapy: *Schistosoma mansoni*, commonly known as a blood fluke, a worm that is responsible for a deadly parasitic disease called schistosomiasis. Humans get infected by this parasite through contact with contaminated water. Adult worms live in their host's blood vessels and lay eggs (hundreds of eggs a day) that lodge in host tissues. It's the eggs that cause symptoms of schistosomiasis; the worms themselves produce chemicals that downregulate the host's immune system response and allow them to control their host environment. The worm's eggs may be a problem, but they are also a point of medical interest: they release a glycoprotein called IPSE, which can easily pass into cells and bind to those cells' DNA. Even better, IPSE can be used as a sort of carrier to transport other molecules, like a DNA sequence, with it.

But if it's the protein that's important, why use worms at all? Well, if you think about it, a good vector must be able to survive the host's immune system with low or no immunogenicity — it shouldn't trigger an immune reaction in the body. It also has to be able to transport the therapeutic sequence to the target cells and to pass through the cell membrane to get to the nucleus where genetic material is stored. *Schistosoma* can survive for ten years or more inside the human body because it has adapted ways of hiding from the host immune system.

Other parasitic worms, called helminths, have been used to treat autoimmune disorders like allergies, asthma, irritable bowel syndrome, and Crohn's disease for the same reason. They trigger the immune system, but in a way that prevents a hyperactive response and reduces inflammation, leading to less severe symptoms. Treatment options aren't great: to get the benefits, you have to be infected by the worms. Basically, you can swallow them or let them crawl through your skin (which kinda makes it clear why Neolution seems to favor implanting the maggot-bots directly).

Of course, the Leda clones are not the only people with the

CLONE CLUB Q&A
In the season two episode "Knowledge of Causes, and Secret Motion of Things," Leekie mentions to Paul that he's developing an artificial womb. Is there any scientific/human health reason why artificial wombs would be useful? Or is it just a made-up sci-fi thing?

Artificial wombs could actually be extremely useful.

Pregnancy and childbirth, even in developed countries, but especially in developing countries, are dangerous and risky and can result in the death of the mother and/or child. There is work being done on ectogenesis — growing a fetus in an artificial environment outside of the body — in the real world, basically with the goal of creating safer options for pregnancy and childbirth.

This technology also has implications for people who have difficulty getting pregnant or who can't get pregnant for reasons like endometriosis, a painful condition where the uterine lining grows in areas outside of the uterus, and for women who don't have a uterus. This is a very real thing — although scientists aren't quite at the stage of having created an actual artificial womb — with positive consequences.

Neolution implant — many Neolutionists willingly have one. Each maggot-bot's therapy is tailored to its host, but other than Aldous Leekie's implant, we don't know what the non-Leda implants were meant to treat or alter. Of course, in "The Redesign of Natural Objects," Detective Duko reveals that there is an ulterior motive to the Neolution implants. He doesn't describe what it might be; he says only that Evie has gotten federal approval for the maggot-bots as medical implants, and that she is intent on making them available to the general public. Based on the dubious ethics of her Brightborn experiments and her obsession with improving humans, it seems that she wants to deliver transgenes to people without their consent and edit their genomes in ways that meet Neolutionist — and not necessarily healthcare — ends. It's really too bad, because its design actually makes the maggot-bot a great tool for personalized medicine . . . if we ignore the whole lethal neurotoxin part.

PERSONALIZED MEDICINE FOCUSES ON PREVENTING AND TREATING DISEASE BY CONSIDERING BIOLOGICAL VARIATIONS AMONG PEOPLE.

Personalized medicine focuses on preventing and treating disease by considering biological variations among people. Variations across individual genomes can affect a person's predisposition to developing specific diseases, but genetic variations can also affect a person's responsiveness to treatment: drug treatments that are super effective for one person might not work as well, or at all, for a second person with the same disease. Some genetic variations can even render a drug toxic to a person. The long lists of possible effects — and possible side effects — that you see on drug labels or flashing in tiny print across your TV screen while being recited at auctioneer's speed during commercials reflect the recognition that individuals respond differently to medications.

For personalized medicine to work, you would want to sequence the individual's genome to determine the gene mutation responsible for the disease. Then you would develop a treatment that is specific to this mutation and to that individual's biology. The maggot-bot's job is two-fold: the gene responsible for the clone disease has not been identified, so its first function is to alter target areas of Sarah's genome to induce the disease. Once the disease has been induced, scientists could track the foreign DNA, thanks to fluorescence, and use this new information to develop a treatment.

Fluorescence is often used by scientists to track foreign DNA, RNA, or proteins because it is visible to the eye, relatively easy to incorporate, and doesn't interfere with the function of the cells. In the case of the maggot-bot, the DNA within the bot to be delivered to the host has a sequence for a fluorescent protein attached to it. When the DNA is incorporated into the host cells and expressed, the fluorescence is also expressed; it can be seen within the cell, indicating that the foreign DNA has been successfully incorporated into the genome. Scientists can then easily collect data and use this information to further develop treatments.

According to Evie, she could tell that Sarah's implant was unsuccessful in isolating the mutation responsible for the clone disease by virtue of Sarah still being alive. That said, we don't know in what ways the bot managed to alter Sarah's genome while it was implanted in her body.

GERMLINE EDITING: PASSING ON SCIENTIFIC PROGRESS TO FUTURE GENERATIONS

The culmination of all the scientific advances to improve the human genome would be to make these genetic changes inheritable, allowing people to pass on their evolved genetics to future generations. Germline editing, a process that involves editing the genome of a person's gametes (that is, sperm or eggs), is unique in that it allows for the edited genome to be passed to their child.

In the case of most of Neolution's advances to this point, the person who has undergone the treatment experiences the modification, but these changes cannot be inherited. The Neolutionists who undergo surgical or cosmetic body modifications may forever have a tail (until Helena chops it off) or a white eye, but if they were to have a child, that child would not have a tail because body modification doesn't affect the person's genes or genome. The same is true of the Neolution gene therapy implant: while Leekie may have been successfully decreasing his chances of developing Alzheimer's, if he were to have a child, he would pass on his genetic risks for Alzheimer's, not the gene therapy decreasing those risks.

The key to the latest generation of the Neolution implant is germline editing. Susan Duncan explains to Cosima when she visits Brightborn that this is the next step in human evolution. The future of the human race — according to Neolution — rests on the ability to pass superiority on from generation to generation.

Genome editing is the process of inserting, deleting, or replacing DNA in the genome, and it is a very real process that scientists all over the world use, albeit only in a laboratory setting. The key to successfully editing the genome is utilizing a nuclease, which is a type of protein that cuts DNA in a designated region. Scientists have used a number of platforms to edit genomes, but the most common and robust platform currently in use is the CRISPR/Cas9 genome editing system. CRISPR (clustered regularly interspaced short palindromic repeats) are sequences of DNA in the bacterial genome that are part of its defense system, working by destroying sequences from viruses. In 2007, a team led by Philippe Horvath used the CRISPR sequences and the Cas9 nuclease to successfully perform targeted genome editing.

In the laboratory, scientists utilize CRISPR sequences to guide the Cas9 protein to specific regions of DNA, which are then cut by the nuclease. This cutting can shut off the target gene by disrupting its sequence. Cas9 proteins that activate target genes have also been developed, and the system has been adapted to allow for modification of the targeted gene as well. In this way, scientists have the ability to achieve extremely precise genetic modification of cells and certain organisms, which may one day be applied to human therapeutics.

Despite the numerous doors CRISPR opens experimentally, there are controversies surrounding this technique. The main issue is where to draw the line on what scientists edit — specifically, whether to edit human embryos. On one hand, editing human embryos can eliminate certain diseases completely, improving the lives of patients; on the other hand, this technology could lead to editing out certain traits that are considered "undesirable," whether

CLONE CLUB Q&A

How long can Helena's embryos be stored? Don't they have to be implanted fairly quickly before cellular division begins?

In season two's "Mingling Its Own Nature with It," Henrik Johanssen, leader of the Prolethean religious cult, holds Helena hostage on his farm, surgically harvests her eggs, and fertilizes them with his own sperm sample. In "To Hound Nature in Her Wanderings," he implants these embryos into Helena's and his daughter Gracie's uteruses and stores the remaining fertilized embryos in a vacuum canister — called a Dewar — filled with liquid nitrogen.

Cell division begins at the moment of fertilization. When eggs are fertilized outside the human body, they are typically implanted into a uterus once the egg has divided enough to become a blastocyst, which is about 100–150 cells. The blastocyst forms about five or six days after fertilization. Since Helena's eggs are in liquid nitrogen, they are frozen in the state they are in. Fertilized eggs can remain frozen and still be successfully revived and implanted after many years.

The only issue with Helena's eggs is that the liquid nitrogen in the canister might run low, and the eggs might begin to thaw. Liquid nitrogen has an extremely low boiling point (-320 °F) which means that it is definitely boiling and evaporating at room temperature. Even the best-sealed Dewars lose liquid nitrogen to evaporation, usually between 0.10–0.30 liters per day. How long a Dewar can keep a sample depends on the volume of the Dewar, its rate of loss (some have better seals than others), and how full of liquid nitrogen it is in the first place. Helena's Dewar isn't very big, and it has already been used for procedures, meaning some liquid nitrogen had already been lost by the time she escaped the Proletheans. (Also, from personal experience:

when you're filling a Dewar with liquid nitrogen, it's difficult to tell just how much you have filled it.) A full, sealed Dewar may advertise a static holding time of 14 weeks or more, but in Helena's case, we'd hazard a guess that it could handle a few weeks, tops. Helena holds on to the canister all through season three, but in the season four episode "From Instinct to Rational Control," Helena learns that her embryos have thawed; she didn't know to add liquid nitrogen. And so, this potential avenue for a treatment for the clone disease ends up buried in the Hendrix backyard.

it be a specific hair color, left-handedness, or the ability to curl your tongue. While many physical traits are not determined simply by a single gene, human genome editing has the potential to pave the way for eugenics and massive alterations to the human gene pool.

These massive alterations can be performed, for example, using gene drives, which is an extension of CRISPR. In Sarah Manning's case study, we looked at Mendelian inheritance and how traits get passed down through generations of pea plants and fruit flies. In any organism that reproduces sexually — so, insects, animals, and most plants — one copy of each chromosome is normally inherited from each parent. This means that two alleles (or versions) of each gene are inherited (and these alleles can be the same or they can be different, like one allele coding for a yellow pea and one coding for a green pea, or two that both encode for green peas). Depending on how these alleles combine, we see different traits. Two yellow alleles will give a yellow pea, a yellow and a green allele will also give a yellow pea, and two green alleles will give a green pea.

Say you had a yellow pea plant and a green plant, and you wanted all of the future generations to have green peas only. With a 3:1 ratio favoring yellow peas, it would take a few generations to produce a handful of pea plants with only green pea alleles, and then you would have to be strict about solely breeding those green-only pea plants together. Random mutations aside, it's pretty easy to do with plants in a controlled environment, but as a rule traits spread slowly through generations and it takes many careful breedings to make a trait prevalent in a population. Imagine trying to control mating among mosquito populations if you were trying to breed in a gene that would make them resistant to the parasite that causes malaria. Imagine trying to do this with a gene in humans — no wonder P.T. Westmorland is losing his patience.

This is where gene drives come in. They change the way in which genes are inherited and how they spread through generations. Basically, it's as if CRISPR technology is being inserted into the organism along with the altered gene, so that the CRISPR is inherited, too. It over-rides any unwanted allele by cutting it out of the genome and pasting in the altered allele instead. Without the CRISPR, the gene could be potentially bred out in the next generation if the organism carrying the gene mates with one that has a different allele that dominates it. When the CRISPR is also inherited, it will go to the specified gene location, check if the altered allele is there, snip out the sequence if it's the wrong one, and replace it with the one we want. In this way, the allele that we want — the altered one — is almost always inherited. In 2015, this was in fact done with mosquitoes to confer immunity to malaria infection and the modified anti-malarial genes were success-fully passed on to over 99% of their offspring.

If Evie Cho was aiming to eugenically change whole populations of people through Brightborn, then gene drives would be the way to go.

THE COLD RIVER INSTITUTE

While Susan arguably has stronger ties to Neolution than Ethan Duncan, who rails that Neolution "killed my Susan," Ethan has been involved in work of equally dubious ethics when it comes to human genetics and eugenics (besides the obvious Project Leda). In season two's "To Hound Nature in Her Wanderings," Sarah embarks on a road trip with Helena in hopes of uncovering the history (and mystery) of Project Leda and finding the "Swan Man," with a photograph of him as her only clue — besides Helena's cryptic explanation that they're headed to Cold River, a "place of screams" that Sarah can't seem to find on any map. Her search takes her to St. James United Church, the building in the photograph. A parishioner recognizes the man in the image as Andrew Peckham (actually Ethan Duncan) and identifies the location as the Cold River Institute. According to the parishioner, the institute had been shut down in the 1970s, but, as it happens, the musty church is the present home to the archives for the Cold River Institute, where Ethan Duncan had been storing its files. (He'd lifted the name Andrew Peckham from one of the stored documents.) Cold River is based on a few real organizations, namely Cold Spring Harbor, home of the Station for Experimental Evolution in 1904. Its original intent was to study heredity and evolution through experimentation with plants and animals, but it soon became a hub for American eugenics — and the Eugenics Record Office from 1910 until 1939. It's worth mentioning that Cold Spring Harbor exists today as a respected research center for biological, genetic, and cancer research, and its early ties with eugenics have long since been severed.

Of the fictional Cold River Institute, the church docent tells Sarah, "The institute was active a lot longer than most people think." Sarah rifles through boxes and folders of documents and photographs, artifacts like a photo of a child with malformed feet and supernumerary toes (a genetic condition known as polydactyly) and another of a baby, labeled "Most Perfect Baby, 1908." Helena was carrying a similar photo in her pocket earlier in the season: look carefully among the candies, broken crayons, and sugar packets that the nurse collected from her at the beginning of "Governed by Sound Reason and True Religion." The *Orphan Black* writers revealed that Maggie Chen stole that photo from Cold River's archives; Helena probably found it in Maggie's storage locker, which we see in "Ipsa Scientia Potestas Est," and decided to keep it.

The photograph points to the bizarre "genetic contests" — with names such as Better Baby, Most Scientific Baby, and Fitter Family — held at the turn of the 20th century. These were exactly what they sound like: families could sign up at state fairs to have their infants and families measured, tested, and generally prodded at by researcher judges who decided which entrants had the most ideal traits and awarded medals that proclaimed "Yea I Have a Godly Heritage." The goal of these contests was to perpetuate and popularize eugenic ideologies.

Eugenics researchers trained at Cold Spring Harbor to learn how to conduct "field work," namely interviewing subjects, taking medical histories, scoring individuals based on their traits, and constructing pedigrees. They also recommended policies for restrictions on immigration, sterilization, and race segregation in efforts to preserve the "best" of the American population.

The connections between Neolution, Project Leda, and eugenics have been established since the beginning of the series. Leekie may sound like a trailblazer when he makes his speech about self-directed evolution in "Variations Under Domestication," but he didn't coin this term: the logo for the Second Annual Congress of Eugenics in the United States in 1921 featured a sprawling tree and the tagline "Eugenics is the self direction of human evolution."

Most of us hear the term eugenics and our minds flash to Nazi Germany and Hitler's use of eugenics to justify his actions. Francis Galton's original definition of eugenics (adapted from the Greek word *eugenes*, meaning "good birth") in 1883 was simply "the conditions under which men of a high type are produced." His focus was on identifying people of "good stock" — that is, healthy, intelligent white men — and encouraging them to reproduce (and to reproduce

more than men of "low type") in order to, in Galton's view, improve the human race. While Dyad scientists would balk at Hitler's eugenics (yes, even the likes of Evie Cho and Susan Duncan) and, to a lesser extent, Galton's definitely racist and classist eugenics, the reality is that they are eugenicists, too, and their experiments are not that far removed from these horrors. Dr. Leekie preaches self-directed evolution and allowing humans to choose their own genetic destinies, but what are the ethical consequences when we stop dealing with individual physical enhancements and start dealing with chosen genetic modifications that get passed down to future generations? Susan Duncan's ultimate ambition was to create a blueprint for superior humans through Project Leda. Ethan Duncan wanted "perfect little girls." Evie Cho wanted to use her maggot-bot technology to eradicate disease and make other changes to alter the genetic makeup of the human population. Where is the tipping point where a noble cause becomes a cull?

Dr. Delphine Cormier, a scientist for Dyad and a monitor for Project Leda — and one of Clone Club's key allies — has been identified as a eugenicist by her peers. Although she defends Neolution as "not eugenical" in season one (it definitely is eugenical, Delphine!), by season three, she

WHERE IS THE TIPPING POINT WHERE A NOBLE CAUSE BECOMES A CULL?

seems to visibly hesitate when Dr. Leekie describes her as a eugenicist It's clear that she did identify as one at some point. Not that identifying as a eugenicist automatically makes you a bad guy: there are many layers and complexities to eugenics. For example, interest in the subject can stem from such ambitions as removing fatal diseases from the human population or the overall improved health and well-being of the human race. One could even get into the study of eugenics based on an interest in the ideas it presents, without any plans to actually carry out these ideas. Delphine's interest in eugenics was married to her work with Dyad, but her changing relationship with Cosima from clinical to personal helps to bring Delphine perspective. While eugenics can be a tool to help people in theory, it should never come at the cost of people's individual rights and autonomy.

"NEOLUTION FOLLOWS THE SCIENCE"

Neolution as a movement has a complex history that has developed parallel to real scientific progress. Beth and Sarah may dismiss it as bullshit, but Neolution takes real scientific breakthroughs as opportunities for human betterment — at least in their narrow view of how humans should be made better — and uses real, existing technologies as means to their ends.

One of the next steps for Neolution, based on the season four finale, "From Dancing Mice to Psychopaths," is to combine Evie Cho's implant technology with Project Leda 2.0: putting implants into a new generation of clones. The goal seems to be to use the clones as test subjects, so that the implant, meant to rapidly trigger genetic change in populations, will be ready and effective once it is distributed globally as a healthcare product. Evie was looking to commodify genetic choice — for the right price-point, consumers could have the genetics they didn't come by naturally. Rachel sees an opportunity to use Evie's technology to pick up where Project Leda left off: to create a new generation of clones with Neolution-desired mutations and to get it right this time. Rachel gets as far as recruiting surrogates, 1,300 of them this time, but she seems to abandon Evie's hit-or-miss technology when she regains access to Kira, who is known to have the treasured Lin28a mutation.

CASE STUDY:
HELENA

ID: Unknown
DOB: March 15, 1984
Birthplace: London, England,
United Kingdom
Status: Alive

SARAH

They told you you were the original,
didn't they? Your people! It's not
true, what they said! We're genetic
identicals.

HELENA

Let me save you.

SARAH

You, me, Beth, the German you
killed! We're all made the same.
Whoever told you different hates
you as much as they hate us.

HELENA

You're wrong.

SARAH

What happened to you, Helena?

(1.04 "Effects of External Conditions")

Helena represents one of the most fascinating cases of any of the
Project Leda clones. Born in secret along with her twin, Sarah, Helena
was sent into hiding in a Ukrainian convent to keep her safe from
Dyad. She spent her time being abused by the nuns and locked in

a basement, eventually leading her to poke out one of the nun's eyeballs. At age eight, she was taken from the nuns by Tomas, a Prolethean who trained the young girl to be an assassin. When we first meet Helena in season one, she is hunting and killing her fellow clones, convinced that she is the original and that Dyad's clone experiment was an abomination.

Part of Helena's success as a clone assassin was her confidence in her motive: she believed that the other clones were abominations because they were born of science and weren't "natural babies." Tomas repeated and repeated this message to her until it became absorbed into her own thoughts and opinions. But how does someone influence another's beliefs and opinions to such a degree?

Thought reform, also known as brainwashing, is the process of systematically altering a person's beliefs and way of thinking, a tactic that has been used in political regimes as well as in extremist religious sects and cults. Another form of "mind control" is known as operant conditioning, which involves using behavioral stimuli and rewards to achieve the desired response from someone. Operant conditioning can be used in extreme cases of abuse, but it is also a method of teaching that can be implemented by parents with young children, or by people with their pets, to enforce and reward specific behaviors.

The concept of thought control or brainwashing is controversial both in reality and in media representations; it's often hard to prove brainwashing because of how intensely the victims will believe the thoughts are their own. Many fictional depictions of brainwashing don't include the key requirements for success: the isolation and absolute dependence of the victim. In the case of Helena, both of these requirements were met when she was with Tomas. Helena interacted only with Tomas and Maggie Chen; otherwise, she was completely alone. Tomas was Helena's only source of food, shelter, and other basic needs for survival, assisting in the success of brainwashing her — and ultimately leading to her success in killing off a number of Leda clones.

CHAPTER 5

"MY POOR, POOR RACHEL"

RACHEL'S BRAIN INJURY

SUSAN
Rachel, you are the experiment.
One day, you may take over. But be
patient. Recover! Everyone needs
a purpose in life. Ours is all in
service of the greater good.

(4.03 "The Stigmata of Progress")

Let's talk about Rachel Duncan's brain. Let's talk about Rachel Duncan's brain and how it was impaled by a projectile pencil stabbed through her eye — not exactly an everyday sort of injury! In the season two finale, "By Means Which Have Never Yet Been Tried," things seem to be going relatively well for Rachel: Sarah Manning has given her unconditional surrender and is being held at Dyad, and her long-thought-dead adoptive father has returned along with an encoded copy of synthetic sequences that, once decoded, should provide insight to Project Leda and the clone disease. Then, of course, her father tells her that she "doesn't deserve him" before drinking a poisoned tea and dying in front of her, and despite Rachel spitting threats and crushing a vial of Kira's blood underfoot, Sarah refuses to give up the key to Ethan's synthetic sequences (which she doesn't actually have). Just when Rachel thought she might have one small victory in telling Sarah Manning, trapped and strapped to a hospital bed, to "enjoy her oophorectomy," everything falls apart. Unbeknownst to Rachel, Cosima and Scott have MacGyvered a clever weapon using a pencil and a fire extinguisher and hidden it in Sarah's room, under a side table. When Sarah pulls the pin, the mechanism forcefully launches a sharpened pencil, impaling Rachel's left eye and sending her crashing to the floor.

First off: could an injury like this happen in real life? Sure. Eyeballs can be punctured easily enough, and the bony socket that forms the only barrier between the eye and the brain is thin (actually it's made up of seven bones and has gaps called fissures and small holes called foramina that let nerves and blood vessels and pencils pass through). What's perhaps more impressive is that the pencil manages to hit Rachel's eye — a small target, even at point-blank range. Not that Sarah is exactly aiming when she pulls the trigger.

> **FIRST OFF: COULD AN INJURY LIKE THIS HAPPEN IN REAL LIFE? SURE.**

When season two ended, viewers were left with many questions: was Rachel dead? If she wasn't, how would such an injury affect her? Would she come back changed? Would she be a new person? Of course, the show's creators had a ton of fun *not* answering these questions during the hiatus before season three. Most of their interviews were sprinkled generously with statements of "*if* she survives"; in one interview, John Fawcett remarked, "Well, you know there's going to be some kind of damage just beyond a missing eyeball," to which Graeme Manson added, "The pencil went pretty deep."

The pencil did go pretty deep. Let's assume that Cosima loaded the launcher with a full-length sharpened 7.5-inch pencil. Based on the amount of pencil still visible after Rachel was shot, we can assume that roughly four inches of pencil penetrated her eyeball. (The authors made comparative measurements to arrive at this assumption. The measures definitely involved pencils and definitely did not involve eyeballs.) The average human eyeball has a transverse (front-to-back) diameter of about one inch, so that's almost three inches of pencil buried in Rachel's brain. We'll come back to this when we discuss what parts of the brain might have been damaged along these three inches.

Have you ever held a human brain in your hands? It's heavier than you might expect (around 1.3 kilograms), and it's *definitely* softer and squishier than you'd expect. It's so soft that just the weight of the brain sitting in your hands is enough for your fingers to leave depressions in the tissue. It's easy to see just how fragile and vulnerable the brain is to injury, even with the skull and cerebrospinal fluid protecting it.

Brain tissue is primarily made up of neuronal cell bodies, often referred to as gray matter, which process signals, and the nerve pathways, or axons, that connect them, referred to as white matter. White matter actually does look white, or at least whitish, if you're looking at a preserved slice of brain; gray matter looks grayish brown. This color becomes visible mostly due to tissue interactions with preservative chemicals, the same chemicals that make brain tissue rubbery and easier to handle so that it can be sliced and poked and probed without falling to mush. The white color is thanks to a fatty sheath of cells called myelin that wraps around axons, insulating them and helping nerve signals travel farther and faster, like rubber-sheathed electrical wires.

> IT'S EASY TO SEE JUST HOW FRAGILE AND VULNERABLE THE BRAIN IS TO INJURY, EVEN WITH THE SKULL AND CEREBROSPINAL FLUID PROTECTING IT.

The brain is often discussed in terms of circuitry, as if the neurons that allow the different areas of the brain to communicate with each other work like wires to directly connect them. That's not exactly the case. Neurons communicate by a combination of electrical and chemical signals, although historically it's been a matter of intense debate whether a nimble electrical spark or a release of chemicals could be truly responsible for something as complex as human thought. People were divided, somewhat whimsically, as "sparks" — who believed strictly in electrical impulse as neuron communication

— and "soups" — who believed in chemical transmission, a sort of neurotransmitter broth pouring between neurons. What can be said today is this: an electrical discharge can be measured when a neuron fires, but chemical neurotransmitters are responsible for the communications between cells. Only as recently as the 1960s did most scientists integrate neurotransmitters into their schema of how different parts of the brain must communicate.

Neurons fire once that signal reaches a certain threshold, like squeezing hard enough on a trigger to fire a gun. Regardless of whether the stimulus that triggers the neuron-firing is weak or strong, the neuron fires in the same way, with the same intensity. To continue the gun analogy, you can't half-fire a gun: it either fires or it doesn't. This is called an all-or-nothing response. If the stimulus is more intense, the neuron may fire faster, or more neurons may fire overall.

But neurons don't fire bullets. The electrical signal that travels down the axons can't jump gaps between neurons, called synapses, like a static shock jumping from your finger to a metal doorknob. That's where neurotransmitters come in. Neurons take the electrical impulse and translate it into a chemical neurotransmitter, depending on what sort of signal it needs to transmit. There are myriad neurotransmitters that use as many different mechanisms to influence neurons, but their main action is the same: activate receptors on neighboring neurons' dendrites and cell bodies. Once neurotransmitters are released into the synapse, they can lock into the appropriate receptors on the next neuron's dendrites and then trigger ion channels to open. Ion channels are like gateways that allow ions like sodium, potassium, or calcium to move into and out of the cell. These ions carry a charge, and the movement of ions changes the neuron's overall charge from a negative state (its usual state) to a positive one. This positive charge rushes down the neuron body to the end of the axon, where, if the threshold is met again, neurotransmitters will be released into the gap between neurons. And then the cycle begins again, to trigger the dendrites of the next neuron.

Nerve tracts are arranged in well-tread pathways for the right signals to activate or inhibit cells in different areas of the brain. Another reason why nerve pathways in the brain are often referred to in terms of wiring and circuits is to describe consistent routes that signals take in response to a stimulus. The Papez circuit, associated with emotional expression and spatial and episodic memory, begins and ends in the hippocampus, which is a jelly roll–shaped deep temporal lobe structure, underneath the cerebral cortex, associated with converting short-term memories to long-term (a fun way to remember this function is to think that you'd "remember a hippo on campus"). It follows a set path through nodes in the cingulate cortex and hypothalamus, each with their own role in emotional response and regulating hormones. The entire circuit is only about 350 millimeters of nerve pathway. Anything that damages any of these circuits in the brain, anything that disrupts neurotransmitters and alters their balanced levels, can alter cell signaling: imbalance or disruption could trigger neurons to fire when they shouldn't, or it could prevent neurons from firing when they should.

You've probably heard of the brain being separated into lobes by function, but even the lobes are mapped out by different cell types — based on their cytoarchitecture, or how the cells look when stained and observed under a microscope — that further define function. These specific areas of the brain were mapped out by German anatomist Korbinian Brodmann in 1909 (and, unsurprisingly, they are known as Brodmann areas). Today, Brodmann areas are useful tools for understanding the functions and relationships between different parts of the brain and allow us to predict what will happen if there is an injury, or lesion, on a specific part of the brain.

That said, often it isn't as simple as knowing that an injury was localized to a part of the head associated with such-and-such Brodmann areas. Since the brain is soft and squishy and easily damaged, the amount and type of damage depends greatly on how the force was applied. The most common head injury is concussion, but there are other common injuries that can have more diffuse effects, such as coup-contrecoup injuries (literally *strike-counterstrike* in French), where you have an injury at the site of impact (the "coup" part) and a secondary injury on the opposite side of the brain where the force of the impact slammed the brain against the inside of the skull (the "contrecoup" part). Sudden acceleration and deceleration (think whiplash) can also cause axons in the brain to stretch and tear in what's known as diffuse axonal injury. If axons (which make up the nerve tracts of white matter) shear, communication between different parts of the brain can be interrupted. And then, of course, there's insidious damage that doesn't even come directly from the injury. Secondary injury can appear over the course of minutes to days after the moment of impact, caused by changes in cranial blood flow and pressures within the skull.

Rachel's injury is what's known as a penetrating brain injury — or open head injury — because an object (the pencil) has passed through not only skin and bone, but also the meninges (the protective layers surrounding the brain) to penetrate brain tissue. Because this type of injury tears through barriers protecting the brain, penetrating injuries are especially prone to infection, especially with a projectile like a pencil, which is wooden and porous and thus a great breeding ground for bacteria, fungi, and a host of other microbes. In Rachel's case, the pencil also crushed the soft tissue of the brain, creating a permanent path of damage along with any bits of bone or other tissue that it might have pushed into the brain. Penetrating head injuries can also damage blood vessels on the brain, leading to a risk of hemorrhaging and swelling, among other side effects.

COSIMA

We're going to do an experiment.
I want you to try to push your
favorite pencil through this
paper. Okay? See if you can do it.
Oops! How come that didn't work, I
wonder? 'Cause you know what? You
need more force.

(2.10 "By Means Which Have Never Yet Been Tried")

CLONE CLUB Q&A

Mrs. S says that women are more likely to survive a gunshot wound. Is that true?

Mrs. S is always prepared for the worst of worst-case scenarios, but when she returns home from Felix's art show, she has a showdown with Ferdinand and he manages to surprise her with a gunshot to the chest. Even while she's dying of a critical injury, she hangs on long enough to sass and distract Ferdinand and find a hidden gun (we did say that she's always prepared — Ferdinand may have found one of her hidden weapons, but he should have known that a pro like Mrs. S would have more than one cached away). Her last words to him?

MRS. S

Did you know, as a woman, I'm
14 percent more likely to
survive a gunshot than a man?
Maybe not this one. But okay,
it's been a good run.

(5.08 "Guillotines Decide")

She's referencing a specific study published in the *Journal of Trauma* in 2010. Researcher Dr. Adil H. Haider and his team at Johns Hopkins University School of Medicine studied over 48,000 patients who were severely injured and had arrived at the emergency room with low blood pressure. Low blood pressure post-injury is a sign of severe blood loss and a strong indicator of traumatic shock. The researchers found that in general women were more likely to survive than men, in particular women between the ages of 13 and 64, even when factors such as type of injury were considered. This is a wide age range, but it's when women are most likely to experience their highest levels of the sex hormones estrogen and progesterone.

Haider suggested that female sex hormones confer some immune-enhancing effect after a traumatic injury. It's unclear how, but they seem to give a boost to cardiovascular, metabolic, and immune reactions. Of course, men do produce estrogen and, conversely, women do produce androgens like testosterone, but the ratios are different and change over time. Men don't produce high enough levels of estrogen and progesterone relative to testosterone to achieve the same benefits. On top of that, other studies have shown that high levels of testosterone might have an adverse effect on a traumatized immune system.

Maybe Mrs. S suspected that it was that hormone-powered boost that bought her enough time to kill Ferdinand, even if it wasn't enough to keep her alive.

Earlier in the episode, Cosima teaches Kira a basic physics lesson using a sharpened pencil and some construction paper: force equals mass times acceleration. Although the pencil's mass doesn't change, increasing the acceleration with which Kira tries to stab at the paper increases the force with which the pencil strikes and punctures the paper. The pencil is more likely to cause damage if it is moving faster.

Likewise, the damage caused by Rachel's injury is dependent on how fast the pencil was going when it entered her eye. Think about the damage caused by bullets: high-velocity projectiles have greater capacity for causing damage as they pass through tissues and can cause more generalized effects like cerebral edema (accumulation of fluid and swelling of the brain, which in turn causes further damage after the initial injury) compared to lower-velocity penetrating injuries from larger objects. People who are injured with non-missile projectiles, like Rachel's pencil, entering the brain at a velocity of less than 100 meters per second tend to have a pretty good prognosis. Higher velocity projectiles will cause more complex injuries and are more likely to be fatal.

What happens to Rachel might ring a bell for anyone who has ever heard the story of Phineas Gage, a bit of neuroscience legend. On September 13, 1848, Gage, then a 25-year-old man working as a foreman on a railroad construction project, was tamping gunpowder down into a blasting hole with an iron rod when he got distracted and turned his head. Accounts differ about what exactly happened next. Did he keep tamping without looking at his work and scrape the rod on the rock? Usually sand is poured into the blasting hole to cover the explosive gunpowder. Did his assistant forget to add sand to the blasting hole, and did Gage slam the rod into straight gunpowder? One thing's for sure: sparks flew, gunpowder ignited, and the tamping rod blasted straight through his skull. It entered through his left cheek, pierced through the front of his brain behind his left eye, and javelined into the ground some hundred feet away, covered in blood and bits of brain. His left frontal lobe was destroyed.

He lived. And although the force of the rod exploding through his skull did knock him onto his back, and he had "a few convulsive motions of the extremities," he never even lost consciousness. In fact, he was reportedly steady enough after a few minutes to climb into a cart by himself to be taken to a doctor. At the time, common opinion was that the frontal lobe was pretty much useless, and the prefrontal lobe was considered "silent" because moderate lesions produced few or no abnormal effects — a view that totally jibed with Phineas Gage walking and talking without much difficulty despite having had his left frontal lobe completely pulped. Frontal lobe injuries are

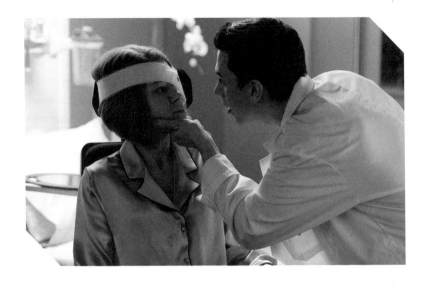

insidious because the behavioral changes they may produce often fall within the range of normal human behavior; these sorts of injuries can go undetected unless the affected person's post-injury behavior is compared to what they were like before. In Gage's case, the doctor decided that his brain must be fine.

These days, the frontal lobe remains somewhat of an enigma — we know that the frontal lobe is tied to many, many functions, from decision making and risk assessment to short-term memory to social behavior and empathy. One of the key categories of functions linked specifically with the frontal lobe is known as executive function. Executive function encompasses skills like working memory, problem-solving, sustained and selective attention, inhibition, response maintenance, and cognitive flexibility — skills that you need to complete complex tasks every day. The frontal lobe is also highly connected to many areas of the brain, so it can be a challenge to pinpoint any of these functions as "belonging to" the frontal lobe and only the frontal lobe. Generally, we can divide the frontal lobe into two main areas: the motor cortex, which largely controls movement and planning movement, and the prefrontal cortex, which is tied to a slew of cognitive and behavioral functions. The prefrontal area, then, is the place where higher thinking happens, and it's thought to be the home of our emotional control and personality.

People who receive prefrontal resections or removal may show a range of behavioral and cognitive signs ranging from driving past stop signs and failure to show concern to rapidly escalating to violence over minor provocation only to quickly revert to a calm state.

How did we figure out what we know now about the frontal lobe? From studying brain injuries like Gage's. Since Gage, there have been a number of similar cases of penetrating brain injuries, enough to justify a fascinating case review titled "Transcranial Brain Injuries Caused by Metal Rods or Pipes Over the Past 150 Years." This 1999 paper includes records of injuries from a wide range of objects, from pipes to crowbars to a heavy metal drill bit, a fishing harpoon, and even an aluminum shaft hunting arrow (in a sort of drunken William Tell incident), many of which caused lesions in similar regions of the brain as those sustained by Phineas Gage and Rachel Duncan. No matter the object causing the injury, the effects were similar: most of those injured were reported to have little or no cognitive deficits. If any did exist, they involved mild difficulties with memory and some challenges with reading, writing, and producing speech, but without any change to comprehension or intellect. Another common result of these injuries was

THESE DAYS, THE FRONTAL LOBE REMAINS SOMEWHAT OF AN ENIGMA — WE KNOW THAT THE FRONTAL LOBE IS TIED TO MANY, MANY FUNCTIONS, FROM DECISION MAKING AND RISK ASSESSMENT TO SHORT-TERM MEMORY TO SOCIAL BEHAVIOR AND EMPATHY.

weakness or loss of use of limbs, most often on the opposite side of the injury. Reports of these changes of limb function ranged from simply using the affected limb less often, a loss of fine motor control, or a tendency to drag (if the affected part was a leg or foot) to more severe paralysis. This loss of movement often resolved itself over the course of months or years with only some residual weakness.

The doctor who treated Phineas Gage may have decided that his patient was fine, especially since Gage was conscious and speaking when he sought medical attention, but after his injury, Phineas Gage was famously "no longer Gage." There isn't a great record of Phineas Gage's life after his injury, and the medical doctors of his time simply wouldn't have been interested in the same information as today's neuroscientists, so the story of what became of Gage is more myth than case study. Some sources cite the extreme: that he was transformed overnight from a pleasant, hardworking young man into a drunken, aggressive, lazy, profane jerkface. Others state that he lost ambition and any sort of money sense (there's a story that one man offered post-injury Gage a large sum of money in exchange for a handful of pebbles, and Gage refused, perceiving regular old pebbles as having a greater inherent value than any amount of money), and that he was emotionally void.

Could Rachel's injury have caused her to experience a Phineas Gage–style change of personality?

Maybe. The part of the prefrontal cortex that sits right behind the eye sockets is known as the orbitofrontal cortex. The orbitofrontal cortex is linked to impulse control, response inhibition, and social behavior. Think of this area as being a great inhibitor — it inhibits impulses, evaluates consequences, and deters inappropriate or socially unacceptable behaviors. When people are drunk, for example, the alcohol inhibits their frontal lobe functions. And what happens to your behaviors when you inhibit the inhibitor? If you've ever been drunk (or have ever seen a drunk person), then you know the effects: you make decisions that you might not otherwise, say things that are socially inappropriate, and act on impulses. In terms of executive function, a person with a frontal lobe lesion may also experience difficulty planning and making decisions, and an inability to alter behavior in response to changing rules. Maybe this is why Rachel cheats against Scott at Agricola in "Community of Dreadful Fear and Hate," although it's far more likely that she simply can't be bothered to learn how to play correctly, even for the sake of keeping up the pretense that she and Scott are enjoying the tabletop game.

Rachel's character is defined by her ruthlessness and impulsivity. She can be a calculating manipulator, but more than once we've seen her emotions get the best of her and push her to extreme and aggressive behaviors. Most often we see her anger barely restrained. An injury to the left orbitofrontal lobe might have left Rachel less

equipped to rein in her impulses. In the season four finale, "From Dancing Mice to Psychopaths," Rachel's rage plays out to its desperate and stabby extreme. Sure, it is possible that without her injury Rachel might still have been pushed by innumerable insults and abuses to turn so violently against Susan Duncan. And, of course, Rachel has been building toward some sort of extreme action against Sarah since they first met, but any past decisions she'd made regarding Sarah seemed more like exacting moves — not the unbridled display of aggression that we instead witness in the season four finale, which sees her striking Sarah repeatedly with her cane, stabbing her in the leg, and shooting at her.

> **IRA**
> You really don't like each other,
> do you?

> **RACHEL**
> She put a pencil in my brain.
> *(4.09 "The Mitigation of Competition")*

Ultimately, we see Rachel stitching up Susan's wounds (and locking her in the very same room that she found herself locked in in the season three finale). Her goal isn't to kill Susan, necessarily: Rachel, as always, seeks control and seeks to prove herself. And to whom does she want to prove herself if not her mother? Pre-injury, it is more likely that Rachel would have quelled this impulse to snatch up a knife and bury it in Susan's stomach and instead have stored her rage for a more calculated next step.

Another common effect of traumatic brain injury is depression, fear, and loss of motivation. Brodmann area 25, also known as the subgenual area — which may lie along the pencil's path through Rachel's brain — is an area that is rich in serotonin transporters, proteins responsible for carrying the neurotransmitter serotonin between neurons. Specifically, these proteins transport serotonin from the cleft between neurons to the presynaptic neuron, recycling it, preventing it from acting on the post-synapse neuron, and thereby ending its effects. Although its primary function in the body is associated with regulating appetite and gut function, serotonin has also been associated with depression. For this reason, these transporter proteins are often targeted by drugs that treat depression and mood disorders. The subgenual area has connections to various parts of the brain, including the hypothalamus and the brainstem, which regulate important functions such as appetite and sleep, as well as the amygdala and the insula, which are tied to emotions, mood, and anxiety. Left frontal lobe injuries may cause depression more often than those to the right frontal lobe.

Characteristics such as depression, fear, and loss of motivation run counter to Rachel's basic character. What would she be like stripped of her basic Rachel-ness? While in the season four finale, we see Rachel unleash her wrath, in the season three episode "Certain Agony of the Battlefield," we see her seemingly vulnerable and sad; it's a normal human response to Felix coming into her room, threatening her, and humiliating her by painting a sloppy eye on the gauze over her eye socket, but it's unusual for Rachel Duncan, who would typically never suffer this sort of treatment. In "From Dancing Mice to Psychopaths," Rachel tells Ferdinand, "I'm not the same . . . my infirmities . . . I don't feel much of anything anymore" in response to his sexual advances. Is this Rachel revealing a change in affect and emotional response as a result of her injury? Or is this just her equivalent to "Not now, dear, I have a headache"?

HOW TO REMOVE A PENCIL FROM YOUR BRAIN

The thing with an injury like Rachel's is that you can't just pull out the pencil without risking further damage. The first thing that doctors would have done once Rachel was discovered on the floor in Sarah's hospital room would have been to stabilize the pencil and perform a neurological assessment, including the Glasgow Coma Scale, which measures elements such as eye opening and verbal and motor response to gauge level of consciousness.

Next, they would image Rachel's head with a computed tomography (CT) scan to determine the location of the pencil in her brain,

and to locate any skull fragments or damaged blood vessels that might pose additional problems. The surgeons would use that information to determine which surgical procedure to use to remove the pencil with the least amount of trauma. An MRI could also be a helpful way to image Rachel's brain, as long as there was no magnetic metal on or in the pencil, since the powerful magnets used in this machine would cause it to shift in the wound and cause more damage. If they were worried about injury to major blood vessels, they might also perform a cerebral angiography, where a contrast agent is injected into the blood so that the blood vessels can be clearly imaged and surveyed for damage.

Rachel would also be closely monitored for any increase in intracranial pressure (pressure inside the skull caused by swelling or edema), since any increase in pressure would indicate a worse prognosis for the injury. The doctors would also monitor for any cerebrospinal fluid leaks, which occur when the protective dura around the brain have been punctured. Cerebrospinal leaks are highly predictive of future infection. Leaky dura should fix themselves, but if they don't, then surgery is necessary to close up any tears. Just in case, Rachel would have been put on antibiotics to prevent any infection. Infection is a huge risk — in fact, in Phineas Gage's case, infection almost did him in. Initially, Gage's condition was said to have been fairly stable for someone who'd had a tamping rod blast through his brain. He welcomed visitors and recognized his family members; however, within a few days a fungal infection blossomed in his wound and caused brain swelling and a dangerous increase in intracranial pressure. This increase in pressure altered his behavior — reportedly, he began raving and demanding someone provide him with his pants so that he might leave his home. As the pressure built, Gage fell into a coma, and his situation was looking grim enough for someone to call for him to be measured for a coffin. He survived only because his medical doctor performed an emergency surgery to reduce the pressure on his brain by draining excess pus and blood through a puncture inside Gage's nose.

Increases in intracranial pressure can also spell an increased risk for seizure activity. That said, seizures are a common occurrence with penetrating brain injuries, affecting up to half of individuals with such injuries; 80% of these seizures occur within the first two years post-injury, but some may not occur for the first time until five or more years later. Rachel does seem to have at least one seizure, when she is visiting Mrs. S's house in "Ruthless in Purpose, and Insidious in Method," although it is subsequently revealed to be a ploy. Rachel and Nealon use the common incidence of seizures to their advantage, giving them the opportunity to steal Ethan Duncan's copy of *The Island of Doctor Moreau*, bring Rachel back to Dyad, spirit her away — replacing her with the kidnapped Krystal Goderitch — and have Nealon tell Delphine that Rachel suffered an intracranial bleed and might not ever recover.

Back to the pencil. Surgically removing it would mean opening up the skull to access the object lodged in the brain. It's nonstandard, but it seems that the Dyad surgeons decided that, since Rachel's left eye most certainly had been destroyed by the pencil, they could access and remove the pencil through her damaged eye socket.

When Rachel finally awakens from her encounter with the pencil, she finds herself at the start of a long road to recovery. Most notably, Rachel no longer has her left eye. The pencil caused irreparable damage to it, and it had to be surgically removed, a procedure known as enucleation. This process involves removal of the eye as well as the muscles that control eye movement and some other nearby tissues to clear out the eye socket. Once the removal is complete, a conformer (a piece of plastic that resembles an eye) is placed to maintain the shape of the eye socket and eyelid, and the eyelid is often sewn shut temporarily to ease the healing process.

Rachel eventually receives her special prosthetic eye, but until then, she covers the socket with bandages and then an eye patch. Covering up the eye socket hides any injury-associated bruising or swelling and limits unnecessary movement of the eyelids that could extend healing and recovery times.

SO . . . WHAT HAPPENED TO RACHEL DUNCAN?

Aside from destroying her eye, the pencil Sarah shot at Rachel penetrated the left frontal lobe of her brain. As we mentioned earlier, the frontal lobe is known as the emotional control center and also plays a big role in personality. Among the other functions involved in the frontal lobe are language, motor function, and memory. Specifically, the left frontal lobe, which is where Rachel was injured, is involved in language-related movement — that is, movements of the face (the lips, tongue, and jaw) required for proper speech. There are two main areas of the brain associated with language: Wernicke's area and Broca's area. Wernicke's area is part of the temporal lobe (often associated with processing auditory information) and is involved in language comprehension. Broca's area sits within the frontal lobe above and behind the left eye and is involved with speech production. It communicates with the temporal lobe, with Wernicke's area, to process sensory information and organizes the string of sounds that a person will speak, making adjustments to the flow of speech. It then sends this direction to the motor area of the frontal lobe, which coordinates the movements of the mouth, lips, and tongue necessary to actually produce the sounds. Many language difficulties have been observed in people with left frontal lobe lesions even when Broca's area itself is spared. The use of language may be abnormal and may exhibit reduced formulations of responses and limited ability for discourse and conversation. It's possible that the pencil directly injured Broca's area in Rachel's brain, but it is more likely that only the pathways that allow for Broca's area to communicate with other areas were damaged.

After her injury, Rachel experiences a specific speech disorder known as aphasia, a communication disorder caused by traumatic brain injury that causes difficulties in speech and language without any loss in intelligence. In general, aphasia refers to a lack of access to language, not a loss of language itself. A person with aphasia knows the words they want to say, but is unable to vocalize and communicate these words due to their brain injury. There are two major types of aphasia, associated with the two language centers: receptive aphasia and expressive aphasia. With receptive aphasia, or Wernicke's aphasia, the affected person is unable to understand and process language appropriately, and though they may speak using proper grammar and vocabulary, their speech lacks meaning, and they don't realize it. With expressive aphasia, or Broca's aphasia, the affected person understands and processes speech properly but has difficulty producing speech. The difficulties associated with this aphasia can include problems coming up with words, substituting words, failing to string words together into a sentence, and making up words. Receptive aphasias are associated with injury to the left temporal lobe while expressive aphasias are associated with the left frontal lobe, in particular the motor cortex, supplementary motor areas, and associated structures below the frontal cortex (for this reason, expressive aphasia is sometimes referred to as motor aphasia). Based on the location of Rachel's injury and the way she clearly understands what everyone is saying to her (outside of her confusion about her injury and waking up without an eye) but can't quite express herself using the appropriate words (especially at the beginning of her recovery), it would seem that her aphasia is expressive.

> A PERSON WITH APHASIA KNOWS THE WORDS THEY WANT TO SAY, BUT IS UNABLE TO VOCALIZE AND COMMUNICATE THESE WORDS DUE TO THEIR BRAIN INJURY.

RACHEL

What happened to my — c — corn
— cob?

(3.01 "The Weight of This Combination")

Aphasia is an acquired disorder, and recovery times differ drastically from patient to patient. In Rachel's case, we see steady recovery with the use of various therapies and healing time. In the season three episode "Formalized, Complex, and Costly," Dr. Nealon tests Rachel's ability to vocalize with a stack of picture cards with common images that Dr. Nealon knows that Rachel can recognize (some of them — such as the Castor two-headed horse symbol — are clearly meant to find out what Rachel knows or remembers). But because of her aphasia, she is not always able to come up with the correct word: for example, she says "knife" when the image is of a key. Word substitutions like these can occur, especially with swelling in the acute phase following injury, but they are less likely in Rachel's form of aphasia than in other forms. A common therapy for aphasia that we do not see in the show is melodic intonation therapy, which assists with word recall by encouraging the patient to hum two alternating notes (one high and one low), then to sing the words or phrases that they find tough to recall using that same melodic pattern (high-low-high-low) while tapping out a rhythm for each syllable. It is thought that this therapy assists with recall because it allows the person to access language in a different way — the area of the brain associated with remembering song lyrics is different from the area of the brain associated with spoken language recall. But just imagine: we could have seen Rachel Duncan *singing*.

One other interesting thing to note about Rachel's impairment is her frequent difficulty with the word "my." "My" is a function word (rather than a content word, like a noun or verb), and function words do tend to be more impaired in expressive aphasia. It's also interesting to see how the writers integrated this difficulty into Rachel's character arc: a very selfish and self-involved person struggles with the possessive — almost as interesting as the content words that *are* impaired for her, such as "control," something Rachel values. As she begins to recover, it becomes easier and easier for her to say the correct word, with less stuttering and hesitation as time goes

on, and when she does have difficulties with a particular word, she becomes more able to select an appropriate substitute that makes semantic sense. As the damaged neurons in her frontal lobe repair, the connections they made prior to her injury reestablish and allow for her to regain her pre-injury abilities.

Another noticeable result of Rachel's encounter with the pencil is her confinement to a wheelchair post-injury. Why would a pencil in the eye cause her to require a wheelchair? Well, frontal lobe lesions can cause abnormal appendicular movements (limb movements) and can affect muscle tone and gait. Generally speaking, this sort of restricted movement is not due to a loss of strength or power in those limbs, but rather a decrease in the initiation of movements as signaled by the frontal lobe. The pencil penetrated the left side of Rachel's brain, and the majority of voluntary body movement is contralateral within the brain, meaning that the right side of the brain controls movement on the left side of the body and vice versa, establishing a "cross-over" of nerve signals. The pencil likely damaged the nerve pathway controlling movement, known as the pyramidal tract, causing the right side of Rachel's body to become paralyzed in a condition called contralateral hemiparesis. It's why she is confined to a wheelchair, why she suddenly goes from being right-handed to being left-handed, and why she must regain the ability to walk with the assistance of crutches throughout season four.

It may seem surprising that Rachel has the ability to graduate from a wheelchair to a stylish cane, but that is the wonder of brain injuries. While effects of traumatic brain injuries are often permanent, hemiparesis, like aphasia, has the potential to decrease over time, especially when coupled with rehabilitation. First Dr. Nealon and then Susan Duncan work with Rachel to regain her ability to walk, write, apply makeup, threaten Sarah's life — you know, the usual. Brain injuries remain a mysterious area of medicine, and each injury and recovery is patient-specific; this leaves Rachel in the dark as to the extent to which she will recover, but she sure seems motivated to return to her pre-injury state.

In "The Redesign of Natural Objects," while making a speech to Ira, Rachel stands up without the aid of her crutches and speaks without any impairment. As Ira notes, "Apparently, passion helps." Expression of emotions can be possible through the right hemisphere of the brain, where our limbic system (or "emotional brain") predominantly lies. In Rachel's case, her right hemisphere would be unaffected by the lesion that is causing her aphasia and hemiparesis. Relatedly, although not seen in Rachel's case, people with expressive aphasia can have an "island" of automatic speech that they can express without difficulty. These islands tend to comprise the sort of speech that you use every day without thinking about it, like saying "thank you" or singing "Happy Birthday." Automatic speech tends to have highly integrated neural pathways, so even if a pathway is damaged by a lesion, there are other pathways by which this type of speech can be accessed and expressed.

AND THAT'S NOT ALL

The brain is key in producing and managing the body's hormones, so injury can sometimes bring unexpected hormonal results. In the

season four episode "The Stigmata of Progress," Ira mocks Rachel by mentioning that her pituitary hormone treatment has been giving her facial hair (a remark at which Susan scoffs, claiming to "hardly see anything"). Pituitary hormone is an umbrella term for any of the eight hormones produced by the pituitary gland, a pea-sized gland at the base of the brain. It's often referred to as the "master gland" because it works like a sort of orchestra conductor, secreting hormones and signaling other glands in the body to produce hormones important to life functions. Of the hormones produced by the pituitary gland, the ones that Ira and Susan are likely most concerned about, in terms of Rachel's recovery, are follicle-stimulating hormone and luteinizing hormone — both important for ovulation and the production of the hormones estrogen and progesterone — and growth hormone.

There are two reasons why Rachel might be receiving this treatment. For one, hypopituitarism (an underactive pituitary gland) after a traumatic brain injury is fairly common, and she might be receiving hormone replacement to mitigate the effects of her injury on her hormone levels. Research has shown that most pituitary imbalances caused by traumatic brain injury are transient and correct themselves within three months to one year after the injury, although there can be lasting effects on the pituitary gland and the hypothalamus (the other "master gland" in the brain) that irreparably affect their hormone production; yet others might suddenly develop these effects over one year after their injury. Growth hormone deficiency is the most common pituitary deficiency post-trauma, but any hormone (and more than one) can be affected by injury. These hormonal imbalances can pose all sorts of problems ranging from the psychological (depression, anxiety, angry outbursts, and other mood disorders) to the cognitive (memory loss, inability to concentrate) to the physical (muscle weakness, a loss of muscle mass, increase in body fat, diabetes, and increased risk of heart disease and stroke).

The second reason: hormones such as progesterone and growth hormone are known to have neuroprotective qualities; that is, they can boost the chance of survival of damaged nerve cells and in some cases even help promote the formation of new nerve tissue. Providing Rachel with hormone treatments will help speed her recovery along. One of the downsides is that, since these hormones exist in a delicate balance in the body, manipulating the levels of one hormone will have effects on the production of other hormones, sometimes leading to undesired effects like facial hair.

We never do see any of this hair growth that Ira teases about, so either her treatments have balanced out within her body, or whoever it is who is maintaining her hair and nails and makeup while she's trapped on the island has also been taking care of any errant facial hair.

At the end of season two, we thought that Rachel Duncan might be dead, but she proves not only to survive her injury, but to become the most powerful we've seen her by the end of season four. Rachel's journey and recovery are a testament to the plasticity of the human brain, its ability to sustain injury and remap and adapt neural pathways to regain function.

CASE STUDY:
RACHEL DUNCAN

ID: 779H41
DOB: March 31, 1984
Birthplace: Cambridge, England, United Kingdom
Status: Alive

NEALON
You'll wake up with a perfectly matched prosthetic eye. You'll be your old self in the mirror again.

(3.08 "Ruthless in Purpose, and Insidious in Method")

After spending most of season three recovering from the pencil incident and graduating from gauze dressing to a stylish eyepatch, Rachel Duncan and Neolution's Dr. Nealon hatch a plot for her to assume the identity of naive clone Krystal Goderitch, who they leave at the Dyad Institute in an induced coma. The end of "Ruthless in Purpose, and Insidious in Method" sees Rachel willingly placing herself in the hands of a German doctor, who Nealon promises will equip her with a leading-edge prosthetic eye.

In the season three finale, "History Yet to Be Written," Rachel awakens in a strange room and discovers that her destroyed left eye has been replaced with a fully functional cybernetic eye. It isn't perfectly matched as Nealon claimed it would be, but over the course of season four she receives injections directly into this new eye to produce a synthetic iris that hides the machinery and matches her natural eye. The iris is made up of layered rings of pigmented tissues connected to muscles that contract and relax to control the size of the pupil, a hole in the center of the iris, which in turn controls how much light is allowed to enter. Technically, Rachel's cyborg eye doesn't even need an iris: the diaphragm inside the camera controls how much light enters through the aperture, much like the iris controls the pupil. As far as synthetic irises go, silicone irises can be surgically implanted through a slit cut into the cornea (the outer lens of the eye), but it's a risky procedure that can cause permanent damage to the eye and loss of vision. In general, iris implant surgery is not performed for people who are sick of their eye color and want a cosmetic change; rather, the exception is made for people who have a condition called aniridia, where the iris is completely missing, either congenitally or thanks to injury.

Cyborg eyes as complex and fully functional as Rachel's do not yet exist, and whole eye transplants have been unsuccessful. The cornea can be transplanted, primarily because it doesn't require a blood supply (corneas for transplant are harvested from the cadavers of organ donors), but the eye as a whole is so complex an organ that it doesn't lend well to transplantation. That said, there are similar technologies that can at least partially restore vision. The most well-known tech is the Argus II implant, which was developed for people with retinitis pigmentosa, a disorder in which the rod receptor cells in the retina slowly degrade and lose the ability to respond to light. These cells translate light into the signals that are sent to the visual centers of the brain. (Cone cells, also in the retina, respond to varying wavelengths of light and allow us to perceive color.) The Argus II technology involves an implant on the retina, which has a receiver antenna and electrodes, and a sunglasses-like headset equipped with a camera and a transmitter. The camera records an image of what is in front of the person, the image is processed into a signal, and that signal is transmitted to the implant in the eye. The implant's electrodes then produce electric impulses that stimulate nerve cells in the retina, which send the impulses through the optic nerve to the visual cortex of the brain to be interpreted as sight.

The Argus II equipment is much bulkier than Rachel's practically invisible bionic eye (users have to wear the headset and transmitter for it to work), and the image it produces is not in color and is nowhere near clear vision. But it allows people who were previously experiencing total blindness to perceive light and simple, high-contrast images like a large white square on a black background. Because it doesn't restore true vision, it also takes some practice for people using the Argus II to learn how to interpret the images they see.

Rachel's eye, however more advanced, seems to use some similar tech. We know that her bionic eye contains a tiny camera that records and processes what she sees, and it would also translate the processed image into electric impulses to be interpreted in her visual cortex. You might be wondering how this works with the normal vision in Rachel's natural right eye. The human brain is already built for binocular vision, which incorporates two images — one from each eye — into a single image. The image that comes from Rachel's cyborg eye is encoded and stimulates the left optic nerve, just like light entering the eyeball and hitting the retina at the back of a real eyeball would. There's a point where the right and left optic nerves connect and partially cross over, known as the optic chiasm, which is critical for that integration that gives you binocular vision. Lesions to the optic nerves at or after the chiasma results in something called hemianopsia, where half of the visual field in each eye is blacked out. Since Rachel's injury occurred before that crossing-over (or decussation) point, her binocular vision and full field of view are preserved. The Argus II relies on an intact retina; it's hard to say how much of Rachel's pencil-punctured eye would have been salvageable for this kind of technology. We pretty much have to assume that Rachel's eye has a direct feed into her optic nerve and perhaps even has a synthetic retina of some sort.

While we have created some amazing technologies to reclaim aspects of vision — from the Argus II for light perception, to glasses that allow colorblind individuals to better perceive color, laser eye surgery, and good ol' glasses and contact lenses for improving visual acuity — we are still a far stretch from developing something as sophisticated and highly functional as Rachel's cybernetic eye.

RACHEL

Look what you've done for me. Look at my eye.

SUSAN

It's almost perfect.

(4.07 "The Antisocialism of Sex")

Rachel's eye may look perfectly matched (or almost perfectly, by Susan Duncan's estimation), but it certainly has its drawbacks. What about the "visions" — messages in the form of swans, whole or decapitated — that Rachel sees through her new eye? It turns out that Rachel's eye does function on a transmitter-receiver model and Westmorland has the ability to project images directly to Rachel's eye, as well as monitor her through her eye, keeping tabs on her from a distance through a tablet. This gives Westmorland the ability to interfere with the implant so that it can pick up secondary input from a separate source at the same time as Rachel's eye-cam. Her brain processes both and integrates them into a single image, complete with the visual artifacts or "glitchy" appearance of a poorly or remotely transmitted feed. This is similar to how some augmented reality applications work. Of course, Rachel discovers this and decides to solve the problem by ripping the eye out of its socket with a broken martini glass. Not exactly the most clear-eyed method, but certainly effective.

CHAPTER 6

"YOUR LITTLE GIRLS ARE DYING"

THE CLONE DISEASE

COSIMA
I'm sick, Delphine.

(1.10 "Endless Forms Most Beautiful")

When pressed to explain why he participated in a project to clone human beings, Ethan Duncan admits that all he wanted were babies — little girls — as a proof of concept for human cloning. Susan Duncan, on the other hand, wanted a baseline for superior human beings. Both of these aims meant that more than a few genetic tweaks were necessary to make the "perfect" Leda and Castor clones a reality. Even at the time of this writing, the precision of genetic modifications still leaves something to be desired, and in the 1970s and '80s, when the Duncans were experimenting with their embryos, the practice was even more imprecise. Imprecision can lead to undesired traits (much like what we see with the "failed" Brightborn baby experiments); for Projects Leda and Castor, these undesired traits manifest in the form of a mysterious and fatal disease.

THE LEDA DISEASE

The first things we learn about Katja Obinger when we meet her in season one are that she's German, she has funky red hair, and she's coughing up blood. "I think I'm dying," she tells Sarah moments before Helena shoots her dead, and Sarah begins to realize that stealing Beth's identity has quickly spiraled into a much more dangerous situation than she could have anticipated. Later in the same season, Cosima begins showing the same symptoms. Sarah's official introduction into Clone Club is Alison's exclamation that "We're clones! We're someone's experiment and they're killing us off!" While this may be a true statement, it misses the important fact that the biggest threat to the clones' safety and survival comes from within.

> **SCOTT**
> What kind of symptoms are we talking about?
>
> **COSIMA**
> Um ... shortness of breath, coughing up blood, that sort of thing.
>
> **SCOTT**
> Who's the subject?
>
> **COSIMA**
> Uh, subjects. And if I told you that, I'd have to kill you.
>
> *(1.05 "Conditions of Existence")*

In "Mingling Its Own Nature with It," Delphine provides Cosima with video diaries that belonged to Jennifer Fitzsimmons, a naive clone brought to Dyad to treat her symptoms. Jennifer was a competitive swimmer, and her diagnosis came with the discovery of polyps on her lungs. According to Delphine, she was one of the first clones to show symptoms. It is unclear whether Dyad's treatments were meant to cure her or to closely observe the progression of her symptoms,

but she apparently received a new form of treatment every week (whether this is meant literally or is simply how Jennifer perceived her treatments is hard to say, but one week isn't a very long time to see if a treatment is having, or will have, a significant curative effect on the disease). One treatment Jennifer received was cyclophosphamide, a powerful chemotherapy drug often used to treat cancers, but also sometimes used to treat autoimmune disorders, such as the two disorders that Cosima mentions: Churg-Strauss syndrome and Wegener's granulomatosis. Both of these disorders specifically affect tissue in small and medium blood vessels, leading to organ damage, particularly in the lungs and kidneys. Cyclophosphamide works by interfering with DNA replication. As Cosima watches Jennifer's video entries, it becomes clear that Jennifer does not get any better, and Cosima rules out these two disorders in favor of a yet-unknown or unclassified autoimmune disease. Jennifer develops sores in her mouth (this may be from her therapies — sores developing on the inside of the mouth, tongue, and gums is a common side effect of chemotherapy, as are hair loss and digestive problems, all symptoms that we see or that Jennifer self-reports in her diaries) and can barely breathe without the assistance of a cannula. In her last video diaries, she describes how she is coughing up pus. After Cosima reaches the end of the video diaries, Delphine reveals that Jennifer died only three days earlier. She and Cosima autopsy Jennifer's body to find her uterus filled with polyps. Cosima identifies these polyps as the most likely cause for the clones' infertility.

Cosima assumes that the clones' infertility is caused by the clone disease, but that isn't quite the full picture. In "Variable and Full of Perturbation," Rachel asks Ethan Duncan about the clones' infertility and demands to know why Sarah is the exception, able to bear children. (Of course, we learn later that Helena is also able to conceive.) Ethan responds that Sarah's fertility is a flaw, and that the Leda clones are barren by design, a control for their experiment: infertility was coded by a synthetic sequence that the Duncans inserted into the Leda genome. But even today, let alone in the 1980s, there are huge challenges to controlling for where modified genes get inserted into a genome; in the case of the Leda clones, when the infertility sequence was inserted, it likely caused a mutation on a different gene, leading to the expression of the Leda disease.

BOTH SARAH AND HELENA ARE PURSUED BY DYAD, PROLETHEANS, THE MILITARY, AND NEOLUTION BECAUSE OF THEIR FERTILITY.

Ethan and Susan inserted the synthetic sequence to make the clones infertile without any other intended consequences. The sequence encodes a protein that functions to degrade the endometrium and prevent full maturation of the ovaries. This same protein is passed from Castor clones to their sexual partners, causing infertility. In the Castor clones, the side effect of the protein is the Castor glitch.

Sarah and Helena are the only known fertile Leda clones. Sarah had Kira, and, upon giving herself up to Dyad in an unconditional surrender and being subjected to swabs and questioning and other invasions in "By Means Which Have Never Yet Been Tried," she reveals that she also had another pregnancy in the past, which she aborted. Helena becomes pregnant with twins in "Things Which Have Never Yet Been Done." Both Sarah and Helena are pursued by Dyad, Proletheans, the military, and Neolution because of their fertility. By season three, we learn that this fertility is due to a complete lack of the infertility protein. This lack could be due to an absence of the infertility sequence, a chance mutation in the sequence that prevents formation of the protein, or an epigenetic silencing of the gene

CLONE CLUB Q&A

What is the science behind Delphine wanting a urine sample from Cosima? What can a urine sample tell us?

Throughout the seasons, Delphine has been monitoring Cosima's health and the progression of her disease through various tests and samples. In season three's "Community of Dreadful Fear and Hate," she visits Cosima's apartment and asks that Cosima provide a urine sample (and Cosima, in turn, asks for Alison to pee in a cup in her stead). There are a few things that medical labs typically check for in a urine sample, such as the presence of glucose, proteins, blood, or ketones. (Ketones are produced when the body breaks down stored fat to make energy when the body isn't producing enough insulin, a hormone that breaks down glucose; their presence in urine usually indicates type 1 diabetes.)

Delphine might be performing protein analysis on urine. Normally proteins aren't found at measurable levels in the urine, but if they are, it is usually indicative of some major health problem that is affecting the kidneys. Blood found in the urine can also be a sign that something serious has happened to affect the kidneys.

Urine can also give insight into how well the kidneys are functioning. We know from a conversation between Rachel and Delphine in season two ("Things Which Have Never Yet Been Done") that the polyps found in Cosima's uterus and lungs have spread to her esophagus and kidneys. Given that knowledge, Delphine must be checking up on Cosima's kidney function to make sure that her condition hasn't worsened.

that prevents transcription of the gene. Whatever the specific cause, it has saved Sarah and Helena from the chance of developing the clone disease, given them their children, and helped Cosima unlock a cure for the disease.

Tony Sawicki is another unique case among the Leda clones for a number of reasons. He is the only known male Leda clone; he's the only clone who knows about the clones and isn't a part of the Clone Club drama; and he's the only known clone currently on hormone replacement therapy (HRT). Tony's testosterone injections affect him in a number of ways, but one interesting and unknown way they may affect him is in regards to the clone disease.

In the case of a non-clone person on HRT, the addition of testosterone to the body will cause menses to stop. Endometrial tissue may continue to build up, but since it is no longer flushed out monthly, it can cause spotting (light bleeding) or pain associated with the built-up tissue. Additionally, testosterone causes the follicles to stop opening and releasing eggs. The eggs are maintained in the ovaries, which can lead to polycystic ovary syndrome (PCOS), characterized by ovaries filled with many cysts.

The buildup of endometrial tissue and failure to release eggs may sound familiar. In season two, Ethan Duncan describes the effects of the clone disease in very similar terms:

ETHAN

It was Susan's sterility concept: degrade the endometrium, prevent ovarian follicles from maturing.

COSIMA

Okay. Horrifying.

(2.09 "Things Which Have Never Yet Been Done")

So, what does this mean for Tony? While his HRT is vital to him as a trans man, it may, in fact, be life-threatening. If the clone disease begins to develop further for him, he may deteriorate much more quickly than any of the other clones due to his testosterone treatments. It's lucky he has gone this long without showing signs of the disease.

THE CASTOR DISEASE

At the very end of season two, it's revealed that Leda isn't an isolated human cloning project. There is a parallel project, Project Castor, with military aims. The Castor and Leda clones are related as siblings would be related, and it becomes clear very early in season three that the Castor clones are also experiencing symptoms of a clone disease.

The first signs that we see of the disease in Castor clone Seth (the mustachioed one) are very different from what we see in Cosima: Seth appears to experience mental "glitches" — sudden and intense difficulties with concentration and reasoning along with poor impulse control — suggesting something of a neurological bent. Paul drops by in "Transitory Sacrifices of Crisis" to perform cognitive tests, strengthening viewers' early suspicions about the disease. Seth and Rudy seem pretty unsurprised by this visit, which suggests that these tests must be done routinely with all Castor clones to monitor their symptoms.

The machine that Paul uses looks something like a Voight-Kampff machine from the 1982 film *Blade Runner*: the test subject is hooked up to sensors, including a camera that tracks the eye, while the person administering the test asks questions. In *Blade Runner*, the test is designed to pick up on biometric indicators (things like galvanic skin response, breathing rate, heart rate, and eye movement) as the subject responds to questions about empathy in order to distinguish humans from androids called replicants. Instead of asking questions about empathy, Paul presents the Castor clones with syllogisms, or logic puzzles, such as "Some maggots are flies, no fly is welcome. Conclusion: no maggots are welcome," for which they simply have to recognize the logic rules of the statements and answer whether the conclusion is true or false based on those rules. (This example would be false, since the maggots that are not flies would be welcome.) As a diagnostic tool, syllogisms have been used to identify schizophrenia and brain damage affecting the prefrontal cortex, an area of the brain attributed to decision making, working memory, attention, social behavior, and personality. Syllogisms also reflect the dual-processing model of reasoning, in which two systems interact to help us form conclusions. The first system is emotional and heuristic, relying on our surface impressions

of information and what belief baggage we bring into the reasoning process; the second system is deliberate analysis of the information, using self-control and working memory to make sure we're getting our conclusions right.

The headset that Paul uses for the test is based on a customary tool for cognitive testing. We often look to pupil dilation as an indicator of cognitive workload, or thinking effort, and to tiny involuntary eye movements known as saccades as indicators of attention. By watching the Castors' eyes as they answer the syllogisms (or struggle to), Paul is able to get a snapshot of the clones' performance in terms of attention, concentration, and stress. Gadget-free, Susan tests Ira's cognitive function in "Ease for Idle Millionaires" by asking him to recall a visual detail from memory — the color of the roof in Italy — which would give her an indication of how his attention, working memory, and visual memory are performing. Visual memory involves a number of different parts of the brain, including the ventral stream, a pathway that is associated with object recognition and which has ties to regions of the brain associated with long-term memory formation and emotion. Although it is a different sort of cognitive test, Susan's approach gets at the same information: if Ira is glitching as a result of brain lesions, his performance will suffer.

Based on the militarization of the Castor clones, it first appears as though the Castor disease symptoms might result from brain augmentation to produce super soldiers. Creating enhanced soldiers is not a novel military goal, though most experiments have historically involved means outside of altering genomes or physically tampering with soldiers' brains. But chemically tampering? There's definitely a history there. Some more famous examples of military experiments include enabling soldiers to go without sleep for 40 hours or more without the ill effects of sleep deprivation, whether by drug use (mostly amphetamines) or by electromagnetism. Or how about transcending normal human vision to see in the infrared range? That's one project that the U.S. Navy took on in World War II, so that sailors could pick out infrared signal lights. They found that specialized vitamin A supplements actually did help to boost infrared sensitivity (vitamin A contains part of a specialized light-sensitive molecule found in light receptors in the eye), but an electronic infrared sensor was soon developed, which made their efforts redundant.

If we had to choose a brain region to tweak to make super soldiers, the anterior cingulate cortex (also known as the ACC) would probably be a safe bet. This area of the brain is associated with functions such as fear learning, pain registration, attention, emotions, empathy, and impulse control — all aspects that have critical roles in military duty. It also plays a major role in human bonding, which might explain the Castors' strangely close relationships. The ACC is also very connected to the amygdalae, two almond-shaped structures in the brain that control rage and aggression behaviors — often called the "fight or flight" part of the brain. Injury to the ACC can trigger fear and aggression; it can also cause issues with attention and judgment. Degeneration in this area is often seen with conditions such as schizophrenia and Alzheimer's disease.

Tweaked or not, these areas of the brain certainly might be connected to the Castor glitch. The symptoms seem to be exacerbated by stress or adrenaline: Seth experiences glitches notably not only during Paul's syllogistic testing (which imposes some stress), but also when he moves to attack Sarah outside of Felix's apartment in "Transitory Sacrifices of Crisis" (where he collapses, and Rudy, realizing that any cure he might uncover would not be helpful to

Seth at this point, chooses to shoot his brother). This jibes with what Dr. Coady does with Helena at the military base: she subjects Helena to stressful situations, from the same syllogistic testing (which Helena is too distracted by the idea of mangoes to complete) to waterboarding (a form of torture), and observes her for any manifestation of clone disease symptoms. It seems that Coady recognizes that stress has been a trigger for the Castor subjects and so assumes that it might be so for the Leda clones as well.

In "Formalized, Complex, and Costly," Cosima and Scott take a rotary saw to Seth's skull in order to get a closer look at his brain and to hopefully determine the cause of his unusual behavior. They prepare slides of his brain tissue and find that, as Scott says, Seth's brain "looks like Swiss cheese" — a strong indicator of prion disease. (The word "prion" was coined in 1982 as a somewhat creative rearrangement of the first letters of the term *proteinaceous infectious particle*. It seems "proin" just didn't quite have the same ring to it.)

AS SCOTT SAYS, SETH'S BRAIN "LOOKS LIKE SWISS CHEESE" — A STRONG INDICATOR OF PRION DISEASE.

Even if you haven't heard of prions before, you've probably heard of some of the diseases known to be caused by them. These are spongiform encephalopathies (that's "spongiform" as in spongy brain tissue — healthy brains are not supposed to look spongy under a microscope the way that Seth's does), such as Creutzfeldt-Jakob disease (CJD), fatal familial insomnia, and kuru in humans and mad cow disease in cattle. Characterized by progressive decline in neurological function, these diseases (at the time of this writing) are 100% incurable and 100% lethal. CJD, the most common of these diseases (but not that common: it's said to affect about one in one million people worldwide every year), is characterized specifically by impaired memory and judgment, changes in personality and behavior, poor motor coordination, and visual disturbances in its early stages (sounds a lot like Seth's "glitches," doesn't it?), all of which become more pronounced, leading to involuntary movements and jerking of the muscles known as myoclonus, and potentially blindness and coma.

Prions are misfolded protein particles that can trigger other, normally folded proteins to also misfold. The way a protein folds determines its function and its ability to interact with other molecules. Think of the protein as a key: if that key is damaged or altered, then it's not going to be able to unlock doors; it loses its most crucial function. As these misfolded proteins build up in the body, they can form amyloids (another thing that Scott notices in Seth's brain tissue slides), which are basically protein clumps that disrupt neural signals and cause cell death. The brain "sponginess" seen in prion disease is caused by the holes that form in the brain tissue where whole areas of cells have died.

Most cases of spongiform encephalopathies are sporadic or acquired through contact with infected brain tissue. It's less common, but prion diseases can also be caused by a mutation to the gene PRNP, which codes for the PrP protein. The jury's still out on the function of the normal form of this protein: some research has suggested that it might have something to do with protecting cells against apoptosis (cell death) or oxidative stress, or with cell signaling, or with copper ions, or with helping the intake of copper ions into cells . . . or with forming and maintaining synapses. It's a mystery, but consensus seems to be that it's supposed to have neuro*protective* functions; but when the protein misbehaves as a prion, it suddenly develops neuro-*degenerative* properties (and we know with certainty that the mutated form of this protein is associated with prion disease).

While Seth's Swiss cheese brain and neurological decline are pretty much textbook for prion disease, there are so many other factors that aren't. Enough to assume that the clone disease isn't caused by a PRNP mutation, but is perhaps a novel mutation to a different gene that causes the protein that it codes for to behave like a prion. Such a mutation could be an unintended effect of gene modifications introduced to produce the clones, such as the infertility sequence.

One of the unusual aspects of the Castor disease is that it's sexually transmissible. Prion diseases are transmissible by blood (people who are infected with a prion disease or who have risk factors for coming into contact with prion diseases are barred from donating blood), but there's no evidence that they're transmissible via semen. We see the effects of Castor disease infection in two non-clone women: Patty, who was sexually assaulted by Seth and Rudy, and Gracie Johanssen, who had sex with Mark. Both women experience hemorrhaging in blood vessels in their eyes, and in "Scarred by Many Past Frustrations," Gracie collapses in a seizure-like episode, very similar to what we've seen of Cosima's clone disease symptoms in earlier seasons. When Gracie goes to Cosima and Scott for testing, they find that Gracie's ovaries have been damaged by an infection, and they notice the presence of the same protein that they found in Seth's Swiss cheese brain.

In "Certain Agony of the Battlefield," Sarah receives a direct transfusion of Rudy's blood, which causes her to develop a high fever and experience fever dreams. (Febrile hallucinations are more likely to occur once core temperature gets over 103 degrees Fahrenheit, Sarah's temperature during her reaction.) Dr. Coady stops a worried Paul from jumping into action, saying that all this is expected, because it is: foreign blood, even from a genetic sibling, contains factors that are alien enough to trigger an immune response. Sarah's fever and racing heart rate are part of her immune system's attempt to rid her body of the foreign blood. There are many potential complications to blood transfusions, but the most common, and most likely what's happening to Sarah, is known as a febrile non-hemolytic transfusion reaction. Sarah's body is producing antibodies against white blood cells in Rudy's donated blood. If Coady doesn't seem too concerned, it's because, of the possible complications, this is considered relatively mild and can be treated with acetaminophen or a similar fever-reducing medicine. Ultimately, however, Sarah does not develop the clone disease like other women infected by Castor. We know from files uncovered by Paul that Coady has conducted these experiments on other Leda clones in the past, such as a Polish clone (identity redacted), and that previous Leda clones did develop the disease or have symptoms of their own disease exacerbated. Cosima explains that the reason why Sarah does not develop the disease, despite having it injected directly into her blood, is that her body does not produce the proteins on which the prion acts.

The Castor and Leda diseases are caused by the same prion — but while the Castor variant of the disease primarily affects brain tissue, like all known prion diseases, the Leda variant affects the endothelium, the tissue that lines blood vessels. Prions are blood-borne, but there is no known prion disease that attacks the endothelium or

gives rise to the symptoms that Cosima and the other symptomatic Leda clones experience. And while sexual dimorphism (visible differences between different sexes of the same organism) is common to many diseases, such as STIs, autoimmune diseases, and some genetic conditions, the degree to which the Castor and Leda diseases differ is pretty extreme. If *Orphan Black* didn't tell us that the Castor and Leda diseases were variations of the same disease, we'd assume that they were unrelated.

Are there other real diseases out there that behave sort of like prion diseases but aren't prion disease? Sure. Most notably, there's a group of conditions known as amyloidosis. Like prion disease, it's caused by a misfolding protein that triggers other proteins to misfold and form amyloids — protein clumps just like the ones that punched holes into Seth's brain. A number of proteins have been linked to amyloidosis and, depending on the protein that is misfolding, the amyloids might form in different tissues, like the kidneys or heart. None of the symptoms for known forms of amyloidosis quite match up with the Leda clone disease, and none are currently known to target endothelium, but amyloidosis at least sets a precedent for prion-like diseases affecting tissues other than the brain. As an interesting aside: the chemotherapy drug that Cosima notes while performing Jennifer Fitzsimmons's autopsy, cyclophosphamide, is used to treat a common form of amyloidosis. Cosima assumed Dyad was using it to treat for autoimmune diseases, but they might have been trying to treat (unsuccessfully) for amyloidosis as well.

Dr. Coady hasn't always known about the nature of the Castor variant: her clones were sick and her initial goal was to find a cure. But one day, Rudy came back to base camp with a woman in tow who was displaying the same symptoms that we see in Gracie and Patty. Upon treating the woman, Coady decided to build a new experiment and had the Castor clones record any sexual encounters and take hair samples to be tested for genetic precursors. Coady realizes that the Castor disease sterilizes women, similar to how the Leda clones are sterile as part of their disease, and works toward weaponizing Castor by using their infection as a form of biological warfare. In season five, we learn that Coady plans to implement the Castor pathogen to target and sterilize people who can't afford Neolution's brand of commercialized evolution. With a significant part of the human population sterilized, Neolution can exert more control over which genetic traits flourish.

There's also the issue of the differences between the symptoms found in Leda clones and those found in women infected by Castor. The end result is the same — infertility — but the progression is way faster in the non-clone women who become infected. Gracie seems to develop symptoms within days, whereas the Leda clones live with their prions into their late 20s before showing symptoms. This difference might be because the prion disease that the Leda clones have is a genetic or familial variant, while Gracie and Patty were infected by a transmissible or acquired variant of the disease. With a prion diseases like Creutzfeldt–Jakob, inherited CJD tends to show a later onset, progress more slowly, and have an overall longer duration than CJD acquired through infection, which generally moves much more quickly (it's always fatal). (Granted, there is also a variant form of CJD that tends to affect younger individuals and which has a longer span of time between first signs of the disease and death.) In terms of known inherited prion diseases, the age of onset does tend to be delayed; they usually show up in the second half of life. For example, one of the more common forms of familial CJD has an average age of onset of approximately 58 years (with a range from 33 to 82 years) and progresses quickly from that point. Generally, most cases of CJD see death within one year of the onset of symptoms.

This is why Coady is so concerned with identifying symptoms early in the Castor clones: the first sign of glitching sets a really short timer against which she has to race to find a cure. By season five, however, she seems to have accepted that none of the Castor clones will be cured of their disease. In "One Fettered Slave," the Castor pathogen is revealed to have been synthesized from Castor samples. The project is terminated, and so is the last remaining Castor, Mark.

CLONE CLUB Q&A

What do you think would have happened if they put a Castor clone's blood into Cosima or Alison instead of Sarah? Cosima is already showing symptoms of the clone disease, so do you think that blood from a Castor clone with the disease would kill her? Would Alison start showing early symptoms, or could she show more advanced ones? Or would it do nothing because the protein is already something that they have in them?

In season three's "Certain Agony of the Battlefield," Dr. Coady transfuses two units of blood from male Castor clone Rudy directly into Sarah. Rudy and Sarah may both be clones from the same original donor, but they come from different cell lines and are not clones of each other. Sarah's immune system responds to the foreign blood and causes a fever, which, as Coady remarks, is not uncommon since Sarah's immune system is likely reacting to Rudy's foreign blood. But generally when a person develops a fever during a blood transfusion, *you stop the transfusion* because a fever could indicate a serious complication. Luckily, Sarah makes it through her fever and fever dreams, and she doesn't show any signs or symptoms of the clone disease.

Both Alison's and Cosima's bodies are already producing the proteins that cause the clone disease symptoms, and Cosima is already showing these symptoms. We can assume that Alison has the symptom-producing sequence because she is infertile (although it is possible that her infertility is caused by a second, unrelated cause). It's possible that transfusing Rudy's blood into either of these clones would exacerbate their clone disease. The symptoms of the clone disease seem to appear and worsen as misfolded proteins are produced and accumulate in the body. The protein accumulation seems to be slow-working, but since Rudy's blood carries the same misfolded protein, it's likely that Cosima would get even worse and that Alison would probably begin to show symptoms.

CHARLOTTE AND THE CLONE DISEASE

SUSAN DUNCAN

Charlotte was cloned from you, you know. "Adam's rib." Four hundred attempts, I believe. Not a viable way to replicate you, but — important nonetheless.

(4.03 "The Stigmata of Progress")

Charlotte Bowles, still just a child, has the unfortunate role of being an experiment within an experiment. She's the only survivor of more than 400 attempts at creating a second generation of Leda clones. Raised by Dyad and Neolution, Charlotte has spent much of her life in a sterile environment while she is poked and prodded by needles and scientists, her fate in the hands of adults who see her for her biology and not for her autonomy and identity.

To top it all off, Charlotte also suffers from the clone disease. For all the other Leda clones plagued by the disease, the symptoms don't manifest until their 20s; Charlotte, however, is roughly eight years old. Charlotte was not cloned from Kendall's original biological material, like the rest of the Leda clones, but from Rachel's biological material, making Charlotte a clone of a clone. Throughout your lifetime, the DNA in your cells undergoes aging — just as your skin becomes wrinkled and your bones become brittle with time, your DNA develops signs of the passage of time. The most evident signs are seen at the ends of the chromosomes, in regions known as telomeres. Telomeres protect the chromosome from degradation and function to maintain chromosome structure. With every replication of the genome during cell division, a small portion of the telomere is lost, shortening the length of this region. This shortening acts as a type of biological clock, determining the age of the cell.

During the cloning process, the nucleus of one of Rachel's cells was used to create the new cloned embryo. Within this nucleus, the chromosomes maintain the telomere length of the source, and therefore the cells maintain the age of the original, and not of the clone. While Charlotte may only be eight years old, her cells are almost 30 years old, causing her to experience the clone disease much earlier than any of the other Leda clones. A tough burden for such a young girl.

You may be wondering why Charlotte has these symptoms when we've never seen them in Rachel. Rachel does have the defect that causes the Leda clone disease, although her only indication so far is her infertility (Alison demonstrates the same lack of symptoms other than infertility). Most likely, Rachel just hasn't shown symptoms yet. That said, there is also a chance that the clone disease has what's known as incomplete penetrance. Penetrance, in genetics, is the proportion of people with a gene (genotype) who also express the trait associated with that gene (phenotype). A condition with complete penetrance will show up in 100% of the people with a given gene. A good example of a condition with incomplete penetrance is familial breast cancer caused by a BRCA1 gene mutation: having that mutation greatly increases an affected person's risk of developing breast cancer (to about 80%) during their lifetime, but there's a small chance that they'll never develop the disease at all. Many people who test positive for the BRCA1/BRCA2 mutation opt to undergo prophylactic surgery to reduce their chances of developing the disease even though they may never develop the disease at all (Angelina Jolie famously underwent a double mastectomy in 2013 for this reason). So far, based on the number of clones that we have seen showing symptoms versus clones that seem to be in the clear, it appears that the clone disease either has very high penetrance or complete penetrance. So, not to be doom-and-gloomy, but your favorite clone who doesn't have the clone disease . . . probably just doesn't have the clone disease *yet*.

There's no way to predict when symptoms may appear for Alison, Rachel, Tony, or Krystal. And, of course, there are very few things that are 100% definite in science, so there is a chance one or two clones might not develop the disease at all, perhaps due to epigenetic silencing, which means that some factor, whether environmental, lifestyle-related, or something else completely, affects how the clone disease is expressed without actually altering the gene itself. That's a bit of a long shot, though. It's more likely that all the clones save for Sarah and Helena would develop the disease.

Dolly the sheep famously showed similar telomere shortening, and this led to speculation that she was aging faster than non-clone sheep by virtue of being cloned from adult cells. But outside of her developing arthritis prematurely (for reasons unknown), Dolly aged more or less like an average sheep.

While it's true that Dolly died earlier than expected, at only six years old (most domestic sheep can be expected to live between 10 and 12 years), it's often rumored that she died of premature old age thanks to her cloned cells — but that wasn't the case. In fact, at the time of this writing, four sheep clones from the same cell line as Dolly are pushing nine years of age, and other than the beginnings of arthritic stiffness in one sheep (to be expected at this age), none of them are showing symptoms of poor health. At nine years old, these sheep have not been tested for telomere shortening; scientists are planning to study their cells once they've died, but don't expect to find anything out of the ordinary. While some studies of cloned animals have found shortened telomeres like Dolly's, still more have found no evidence of difference in aging between cells from clone and non-clone animals. It's been suggested that these instances of shortened telomeres are an effect of the methods used to create and

implant the cloned embryos. Inefficient methods may interrupt the processes that form telomeres, causing the embryos to form "old" cells instead of properly reprogrammed ones. Scientists struggled to create Charlotte, so it would not be surprising to find that their methods contributed to her premature sickness.

If Dolly didn't die of old age, what happened? According to scientists at the Roslin Institute, where Dolly was born, she died of a disease known as jaagsiekte, or ovine pulmonary adenocarcinoma. While the cause of Dolly's disease is very different from what is causing the clone disease, there are some strong parallels between Dolly's symptoms and the symptoms we see in Jennifer and Cosima. The term *jaagsiekte* is derived from Afrikaans words combined

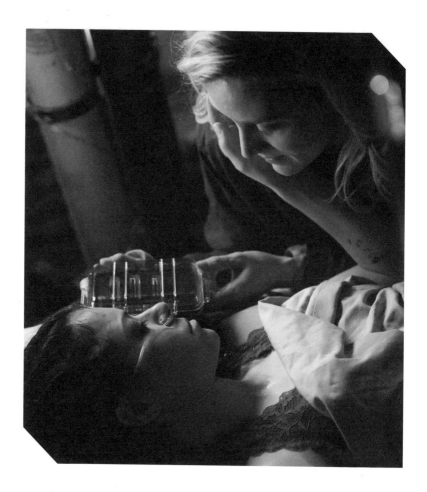

to mean, basically, "chasing sickness," so named because the most obvious symptom is shortness of breath and difficulty breathing, as if the sheep had just been chased by a predator. This is a contagious and incurable disease among sheep and goats that is caused by a virus. In January 2000, one of the other cloned sheep at the Roslin Institute, named Cedric, died, and an investigation revealed that his lungs were filled with tumors caused by the virus. It was unknown if Dolly had been infected; just in case, they isolated Dolly from the other cloned sheep and hoped that she was infection-free. A few months later, Morag, another cloned sheep who had lived closely with Dolly for many years, also died of the disease. Chances were slim that Dolly hadn't been infected. Once Dolly gave birth to lambs that also had the disease, it was clear that Dolly must have the disease. In February 2003, an animal care worker noticed that Dolly had started coughing (but not bloody coughs like Leda clones; jaagsiekte tumors in the lungs cause a foamy white fluid, and lots of it). They investigated with a CT scan to confirm that the suspected tumors were growing in her lungs. Since the disease is incurable, and the veterinary team did not want Dolly to suffer, a decision was made to administer a lethal dose of anesthetic.

HOW TO CURE A CLONE

COSIMA
You know, nobody's got any idea. We're just poking at things with sticks.

(2.06 *"To Hound Nature in Her Wanderings"*)

When we first meet her, Cosima Niehaus is already deeply invested in understanding her own biology. When she realizes that she is sick, she is consumed with finding a cure. Over time we've learned a lot about the clone disease, from Cosima's attempts at treating herself, from other clones who have died before, and from what information Sarah uncovers having been thrown headfirst down the rabbit hole after barely coming to terms with the fact that she's a clone. The clone disease is novel: standard treatments for cancers and autoimmune diseases are ineffective. It behaves somewhat like prion diseases, which are currently incurable.

Where do you begin to search for a cure or even a treatment?

KIRA'S BONE MARROW

Once Sarah discovers that Cosima is showing signs of the clone disease, she wants to do anything she can to help save Cosima's life, including involving her daughter Kira. Although Kira was perfectly okay with

pulling out her own tooth to provide Cosima with some of her stem cells, it's a much riskier decision to follow through with extracting Kira's bone marrow to provide Cosima with transplant material, which she does in "Things Which Have Never Yet Been Done."

Kira is the only viable option for Cosima as a source for hematopoietic stem cells (blood and immune system stem cells that reside in the bone marrow). In order to receive a transplant from someone, it is required that they are a good match; that is, the donor and the recipient must have the same human leukocyte antigen (HLA) markers. Because these markers (which are proteins used by the immune system to identify your own cells versus foreign cells) are inherited, oftentimes a relative is the best match for a transplant.

The majority of the Leda clones wouldn't make for a successful transplant because they carry the same DNA encoding of the clone disease, and therefore their stem cells would make no difference for Cosima. While Sarah and Helena are the only identified Leda clones with a difference in their genome (because they do not make the protein that leads to the clone disease), at the time Cosima needed the bone marrow transplant, not enough was known about the differences in Sarah's genome versus the other clones to make her a viable donor. This left Kira, whose genome was 50% from Leda and 50% from Cal — similar enough to make it very unlikely that Cosima's body would reject the transplant, but dissimilar enough to ensure that the stem cells would help and not hurt Cosima's immune system.

Despite the willingness of Sarah and Kira to follow through with the bone marrow transplant, it is only a temporary solution to the severe autoimmune disease that Cosima is suffering through. Eventually, Cosima's defective immune cells will overpopulate Kira's transplanted cells, and Cosima will continue to get worse. The transplant contains a finite amount of stem cells, and Kira would need a lot of time to recover from the donation before she would be able to donate again. This is why Cosima only gets one transplant from Kira: it's just enough to help her along while she struggles to find a permanent cure.

HELENA'S BABIES

During Helena's stay at the Prolethean Ranch, Henrik decides to take advantage of her fertility by using IVF to impregnate her with her eggs fertilized by his own sperm. Moving past all of the *incredible* ethical breaches Henrik makes in this process, the result is that Helena carries twins, who she is excited (and a little apprehensive) to welcome to the world.

Henrik also impregnated his daughter Gracie with these embryos, although she miscarried after contracting the Castor disease from Mark. Henrik seemed to be preparing to impregnate all of the Prolethean women based on the fact that Helena escaped from the burning ranch with a liquid nitrogen tank full of her "babies." Although these embryos ended up buried in the Hendrixes' yard after Helena accidentally caused them to thaw, they had a lot of biological potential.

Early embryos just past the stage of fertilization are basically just balls of stem cells. The early cells are all pluripotent cells, capable of giving rise to any type of cell in the body — similar to the cells taken from Kira's bone marrow to temporarily help Cosima with her illness. Helena's "babies" could've been used to treat Cosima, if Helena would have ever agreed to give them up. These embryos could also have been used to develop a stable cell line to use in the laboratory, which would've allowed Cosima and Scott to study the clone disease and perhaps allowed them to reach the cure on their own.

> **HELENA**
> Little science babies, forgive me.
> I did not know to feed you liquid
> nitrogens.
>
> *(4.04 "From Instinct to Rational Control")*

If Cosima still needed a source of stem cells, Helena could have been of assistance. Within the umbilical cord attaching a fetus to its mother are stem cells. Many parents choose to have these umbilical stem cells frozen and saved when their children are born because these cells could later be used in the case of the child or a sibling developing an autoimmune disease for which stem cell treatment would be beneficial. These cells, having come from the child, have a greater success rate of being the necessary match for treatment than cells from an unrelated (or less closely related) donor. In season five, Westmorland also wants cells from Helena's babies' cord blood, but he wants them specifically because they hold a desirable mutation. That mutation, which Kira also has, doesn't provide a cure for the clones' disease.

ABEL JOHANSSEN

While Charlotte Bowles represents the second generation of the Leda clones, the Castor clones lack a comparable attempt at maintaining their legacy into the future. Interestingly enough, however, Sarah, with the help of Castor clone Mark, discovers that Henrik Johanssen, the Dyad lab technician turned Prolethean experimenter, has made his own secret attempt at propagating the Castor genome.

In season three's "Newer Elements of Our Defense," Sarah uncovers the grave of an infant, Abel Johanssen, along with Henrik's lab notebooks, subsequently unveiling Henrik's attempt at a second generation of the Castor clones. The notebooks reveal that Henrik got his hands on the original Castor biological material, cloned it, and had his wife, Bonnie, carry the cloned embryo to term. Although

Abel was a live birth, unlike all but one of the second generation of Leda clones, he unfortunately didn't live long, which Sarah saw evidence of in his tiny coffin and even tinier skeleton.

Why would Henrik have gone through the grueling process of attempting to repeat the cloning of the Castor genome? For one, having a living child with the Castor genome would have given Henrik constant access to Castor biological material, which would have been of interest to him both as a scientist and as an antagonist to the original cloning project. For another, if Henrik was aware of the Castor glitch, Abel would have provided Henrik with a substantial source of the original genome without the added synthetic sequences, allowing him to study and possibly develop a cure for the Castor disease.

The thing about sequencing Abel's genome is that he's dead. Very, very dead. DNA is not exactly a hardy molecule, and it's thought to weather poorly. Depending on a body's exposure to moisture and the elements, DNA can become damaged and degraded. Samples can also be contaminated by the DNA of microbes that have been feasting on the body. Abel's skeleton was in a casket, which does help to protect it somewhat, and the bones seemed to be in good shape. Luckily for our clones, bones and teeth are pretty much hot spots for getting a decent sample (like how Kira's stem cells for Cosima's treatments were sampled from the pulp of a baby tooth and from bone marrow in her hip).

In the past, scientists have successfully sequenced genomes from the bones of remains much older than Abel's. In 2010, a full Denisovan (an extinct species of human) genome was sequenced from nuclear DNA pulled from a single finger bone; the finger's owner is thought to have walked this planet some 41,000 years ago. More recently, in 2014, scientists sequenced the genome from a one-year-old Pleistocene (a.k.a. Ice Age!) human from the oldest-known burial site in North America, the Anzick burial site in western Montana. The tools and tech definitely exist.

If a good enough sample of DNA could have been extracted from Abel, then Castor could have been one step closer to a cure. This avenue goes largely unexplored in *Orphan Black*, perhaps because Sarah and the others are much more intent on finding the original clone genomes, which would be a much more viable source for creating a cure. And then it's too late: Abel's remains are destroyed in a lab explosion in "Certain Agony of the Battlefield."

ETHAN DUNCAN'S BOOK

COSIMA

Whoa, *The Island of Doctor Moreau.*

KIRA

It's about a man who makes monsters.

COSIMA

I know, I love this book.

KIRA

(whispers)

It's special.

(2.10 "By Means Which Have Never Yet Been Tried")

In season two, Ethan Duncan gives Kira a copy of *The Island of Doctor Moreau* that contains images and ciphers. From what we can make out from the images, there seems to be information about molecular structure and function, probably of a protein either encoded by or affected by the synthetic sequence that leads to the clone disease. These images include a diagram of a cell and a membrane receptor, possibly revealing information about a specific pathway triggered by a membrane receptor and resulting in specific gene expression. Molecular structures are written out on the top of one of the book's pages, as well as what appears to be an amino acid sequence of a protein. A third image references molecular bonding, possibly alluding to secondary or tertiary protein structure, or protein interactions. Finally, we very clearly see a molecular structure of a protein, perhaps the key to the clone illness or at least a possible player in the disease.

Before Ethan's death, he provides Cosima and Scott with at least one cipher to decode, but there are more in the book, likely encoding each synthetic sequence in the Leda genome. The single cipher is not enough for Scott and Cosima to move forward in developing a treatment, and so season three finds them enlisting Rachel's help. Rachel does decode a section of the book to reveal a poem and a prisoner number that leads Sarah to Kendall Malone, the source of the original genomes for the cloning projects, before Rachel and Dr. Nealon steal the book. The book could have been a promising resource: once the appropriate synthetic sequence was located, Cosima could have induced overexpression of that sequence via gene therapy, effectively silencing the infertility sequence and curing herself of the clone disease.

Ethan's book is also likely a source for other important information: information that gives more insight into the clone's biology, and information that would allow a new generation of clones to be made without the problems faced by Charlotte's generation.

GENE THERAPY

In the *Helsinki* comic issue #4 ("False Positive Error"), self-aware Finnish clone Veera Suominen meets clones who have been held for experimentation by Dr. Dmitri Volkov. These comics take place circa 2001, so these are probably the earliest known sick clones. Volkov was part of projects at the Cold River Institute in the 1970s (around the same time that Ethan Duncan would have been working there), and his research focused on mapping genetic diseases in isolated groups. When we encounter him in the comics, he has at least nine Leda clones in his lab receiving mysterious treatments. Veera steals a hard drive with information about injections, stem cells, target genes, and cancer, suggesting that Volkov was using the clones for gene therapy experiments. We also meet Jade, one of Volkov's subjects, who is clearly ill from the Leda disease, from Volkov's treatments, or both. Like other clones, she has a mysterious scar from an incision behind her ear (in season one, Paul indicates that Beth had a scar like this when he notices that Sarah does not have one . . . could this be a link to Dyad and their treatments?). While trying to escape Volkov's lab, Jade and Veera find themselves in a room full of malformed fetuses in jars. Failed cloning attempts preserved for testing?

While Volkov is one of the bad guys, his area of the study — gene therapy — is the key to curing the clone disease. Gene therapy is a term that is becoming more and more common in popular science, but what exactly does it entail? Simply put, gene therapy involves curing a disease at the molecular level by either adding to, deleting from, or altering a genetic sequence to overcome disease. In the case of the Leda clones, it first requires a comparison of Cosima's genome to Kendall Malone's genome to identify the specific sequence causing the clone disease, and then developing gene therapy based around this sequence in order to eliminate or prevent the formation of the problematic protein.

This technique is not one of science fiction, although its success hasn't yet lived up to its potential in reality. The first ever use of gene therapy in humans was in 1990: doctors used a virus to help treat two patients (a four-year-old and a nine-year-old) suffering from adenosine deaminase (ADA) deficiency, which affects the immune system. These patients went on to live normal lives after their treatment, and the treatment has been used to treat ADA deficiency in newborns successfully; this is one of the better cases in the history of gene therapy.

The success of the ADA gene therapy led to a worldwide proliferation of interest in the technique, with new methods and trials appearing around the globe. Gene therapy excitement continued to grow steadily until 1999, when the worst-case scenario for a gene therapy trial finally happened: a patient in the trial died. Among the patients in a clinical trial for ornithine transcarbamylase (OTC) deficiency at the University of Pennsylvania was 18-year-old Jesse Gelsinger, whose immune system reacted swiftly to the high dose of viral vector, causing him to die of multi-organ failure only four days after his treatment.

Jesse Gelsinger was the first patient in such a trial whose death could be directly linked to the gene therapy. Following Gelsinger's death, multiple initiatives, both legislative and scientific, put a hold on gene therapy in order to improve the safety of the method. Consequently, the interest and enthusiasm for gene therapy by both doctors and patients began to wane due to the potential risks.

Patient death wasn't the only risk haunting the growth of gene

CLONE CLUB Q&A

Do you think that the Leda clones can also spread their clone disease sexually, like the Castor clones do? If so, why aren't the Leda clones' sexual partners (Donnie, Cal, Paul, Vic, Delphine, and so on) sick?

The Castor clones and the Leda clones are both affected by a mysterious illness, but the Castor clones are able to pass it on to other people like a sexually transmitted disease. We see the effects of the Castor disease on Patty and on Gracie Johanssen. Dr. Coady and her military team seem well aware of the Castors' ability to infect others, and they even have the clones recording all of their sexual encounters in sinister "little black books." When Paul uncovers documents detailing the Castor disease and records of what's happened to those who have been infected, he suggests that Dr. Coady is using the clone disease as a weapon, a sort of experiment in biological warfare. The question then becomes: was this ability to transmit the clone disease intended by military scientists all along, or was it something that they noticed and chose to exploit?

If it's the former, then it's unlikely that the Leda clones can also transmit their disease — because disease-spreading clones don't seem to fall into Dyad's area of academic interest. If it's the latter, then why aren't any of the Leda clones' lovers sick?

Our best guess is that if the Leda version of the clone disease can be passed on to others, then it must have a different latent period than the Castor version of the disease. A latent period is the span of time between initial infection and when an infected person begins to show symptoms. Gracie and Patty both develop high fevers and bloodshot eyes, and their ovaries begin to deteriorate, over a relatively short period after having sex with a Castor clone (it's hard to pinpoint an exact timeline, but it appears that we're looking at a matter of weeks). Some of the documents that Paul finds state that other sexual partners have died from the disease. Many of the Leda clones have long-term sexual partners (Donnie) or have recently seen sexual partners from years past (Cal for Sarah, Jason for Alison), and these partners show no signs of infection. If they do carry the disease, then they might finally start to show symptoms years and years after the original infection, or they might never show symptoms at all.

therapy. Another problem became apparent in association with gene insertion: when the vector containing the therapeutic sequence is given to the patient, the goal is to have the sequence insert itself into the patient's genome, so that it can be properly incorporated as part of the DNA. This incorporation process, when using common gene therapy vectors, is completely random; that is, the therapeutic sequence will insert at a random site within the genome.

Only 3% of DNA represents genes that encode proteins, so there is a very great chance that the insertion process won't interfere with any other gene. However, if the gene therapy inserts itself within a gene or within an important regulatory region of the genome, there can be huge consequences. Such was the case in a number of severe combined immunodeficiency (SCID) trials, in which patients developed leukemia after the viral vector inserted itself in positions in the genome that caused either cancer-related genes to become activated or cancer-preventative genes to be silenced. The risk of cancer and other long-term effects has further slowed down both the push for and the progress of gene therapy. Nevertheless, in the case of the clone disease, gene therapy presents itself as the most promising cure, and the personalized and specific nature of the disease makes many of the pitfalls of gene therapy irrelevant.

A crucial component to gene therapy success is choosing the right vector to deliver the therapy. Viral vectors harness the ability of viruses to infect cells and deliver genetic material into these cells, but they are modified to prevent replication of the pathogenic viral genome. There are a number of factors required in order to find the best viral vector for a specific gene therapy. First, the viral vector must be sufficiently safe, and any viral genes that might replicate and interact with the host genome in ways that might cause disease must be deleted. The vector must also have low toxicity (little to no side effects once the virus infects the patient's cells) and high stability (inability to rearrange the genome or degrade within cells). Last — and very importantly — for gene therapy, the viral vector must be able to infect the cells of interest; that is, the cells that are affected by the disease the therapy is targeting.

> **SCOTT**
> One of our big problems is we can't get Cosima's cell population to grow past the lag phase.

> **COSIMA**
> And I'm not even sure we're using the right viruses.

> **SUSAN**
> You've used both retro and adeno-viruses. Have you tried naked DNA?

> **COSIMA**
> Have you seen our lab? We have major trouble controlling contamination.

> **SUSAN**
> Still, it's remarkable what you've achieved with so little.

> *(4.08 "The Redesign of Natural Objects")*

Viruses are the most common vectors for delivering genes into cells (for either transient or permanent expression of those genes), and different viruses have different advantages and disadvantages. For example, of the vectors that Cosima and Scott attempted, adenovirus is generally a pretty safe bet as it can deliver to a wide variety of human cells and has high gene transfer rates, but it can also have high immunogenicity (it triggers immune response). Retroviruses provide pretty stable gene expression, but there's a possibility of the virus causing an unwanted mutation because this type of virus permanently integrates its genome into the host's genome (whereas adenovirus is nonintegrating). The failure to find a vector is scary for Cosima, whose health continues to decline. As Cosima explains to a worried Mrs. S in "Transgressive Border Crossing," because she is going to be the experimental subject for testing the gene therapy, she cannot use any other form of treatment at this time, not even another bone marrow transplant from Kira. When testing the effects of a new scientific therapy in a laboratory, the experimental conditions require subjects who are not experiencing any other form of treatment. This way, the scientists are able to discern what the effects of the new therapy are on the patient, eliminating any effects from combining therapies. For Cosima, this means she can't do anything to treat her illness while testing the gene therapy in order to see if the therapy is functional.

In "The Redesign of Natural Objects," Susan Duncan suggests using naked DNA, which is probably the simplest example of a nonviral vector. Naked DNA has been modified to remove the proteins that usually surround it, and it isn't packaged into anything (for example, a virus) to be delivered into a cell. Naked DNA is often less effective at entering cells than viruses, but it's not specific to any cell type, like viral vectors often are, and it won't trigger an immune response. Unfortunately, as Cosima mentions, this hasn't been an option because their makeshift lab under a comic shop isn't exactly ideal for controlling contamination.

The most important part of this episode is Cosima's revelation and subsequent plan for continuing their research on curing the clone disease. In order to test out gene therapy methods, Cosima and Scott need a stable, renewable source of stem cells to experiment with. Kira may be able to donate stem cells from her teeth or bone marrow, but her sources are finite and, as Cosima mentions, Cal's genetic contribution to Kira's genome makes her cells less than ideal for testing. Ideally, they would want to use cells that are as close to the original Leda genome as possible.

Embryonic stem cells (ESCs) are equivalent to the multipotent cells of the early embryo: they can become almost any type of cell, which enables testing of any cell type of choice, and the cells can self-renew, meaning they can be kept in culture for long periods of time, removing the need for a constant source of new cells. ESCs are obtained from cells that form the inner cell mass (ICM) of the blastocyst. The blastocyst is an early stage of embryonic development, before implantation into the uterine wall, in which there is a single layer of cells surrounding a mostly empty cavity that contains the ICM that becomes the embryo.

Cosima's plan is to generate a blastocyst and harvest the ICM to create a culture of ESCs that she can then use to continue her research. Since the key to this plan is obtaining a blastocyst that is as similar as possible to Kendall's cells, they would have to provide a sample that is as close to the original's genome as possible without actually having access to the original (since Kendall was killed, and Sarah destroyed any existing samples with bleach). A combination of the Leda and Castor genomes can be achieved by fertilizing an egg from a Leda clone with the sperm of a Castor clone and allowing this fertilized egg to develop into a blastocyst. This is where Sarah and Ira come in, because Sarah's eggs are fertile and can therefore successfully generate a blastocyst after fertilization. Sarah, of course, isn't too keen on donating her eggs to be combined with Castor sperm, but she agrees with the plan, if only because it may be the only way to save Cosima without her resorting to even more dangerous methods — like implanting a maggot-bot in her cheek in hopes that the gene therapy that Brightborn programmed it to

COSIMA ISOLATES STEM CELLS FROM INSIDE THE BLASTOCYST AND GROWS THEM IN CULTURE, DEVELOPING A WORKING LINE OF SELF-RENEWING CELLS WITH THE COMBINED LEDA AND CASTOR GENOMES.

deliver might cure her, and not kill her, like it almost did in "The Antisocialism of Sex."

Transported to the island to work in Susan Duncan's lab, which is much more suited for these sorts of experiments than the Rabbit Hole lab, Cosima isolates stem cells from inside the blastocyst and grows them in culture, developing a working line of self-renewing cells with the combined Leda and Castor genomes. She then needs to test the vectors with the inserted gene therapy sequence on the cells. In this case, Cosima exposes the cells to the vector and looks for successful uptake of the vector into the cells, incorporation of the sequence into the cell genome, and survival of the cells after exposure to the vector.

CLONE CLUB Q&A
What was Coady doing to Parsons?

For most of season three, Helena is held captive in a holding cell on a remote military base. In the episode "Newer Elements of Our Defense," she escapes to explore the compound and discovers a Castor clone named Parsons, who has had the top of his skull removed and electrodes inserted directly into his brain. And he's conscious.

Parson's brain appears to be discolored; most likely it's dead tissue, no longer functional thanks to the clone disease. This dead tissue is a likely source of the "glitching" seen in the Castor clones. The electrodes are there either to get a readout of any brain activity still present in those damaged regions or as an attempt to restore some brain activity by shutting off certain parts of the brain. For conditions such as Tourette syndrome and Parkinson's disease, electrodes can be inserted into the brain in a process known as deep brain stimulation. The implanted electrodes send controlled pulses of electricity to target areas of the brain, to stimulate them and to decrease symptoms. Coady could have been trying to turn off the glitching symptom associated with the Castor clone disease.

Cosima's tests are successful in Susan's lab, and she manages to find a vector and sequence combination that keeps the cells alive after exposure and uptake. The therapy is promising: the cells even exhibit a self-renewal rate (the rate of cell division and therefore doubling) that indicates to Susan that the cell line could be used to continue the Leda human cloning project. After a harrowing escape from Susan and Rachel with the cure, only to be found out inside the clinic at Revival, Cosima finally receives an injection of the cure into her uterine lining (the tissue of origin of the disease). By "Let the Children & Childbearers Toil," Cosima and Charlotte (who also receives the cure) discuss how Cosima's breathing has improved, indicating a decrease in the polyps in her lungs.

By the end of season five, the cure is deemed a success, saving Cosima's and Charlotte's lives, allowing them to move on to cure other clones. Rachel is treated, and Cosima and Delphine begin their trip around the globe, hoping to cure all 274 of the Leda clones.

CLONE CLUB Q&A
Do you think the clone disease can be treated like cancer, or is that too risky?

Bone marrow transplants are a common treatment for cancers such as leukemia and myeloma. Obliterating the patient's bone marrow with chemotherapy or radiation and then performing a bone marrow transplant can be an effective way to remove the cancerous cells and effectively reboot the immune system. So, in that way, the clone disease is technically already being treated like some cancers.

Because the clone disease is an autoimmune disease, simple chemotherapy and radiation wouldn't work without the bone marrow transplant. The chemotherapy would target the blood cells and polyps, perhaps reducing the effects in the short term. But once the bone marrow replenished itself, the immune system would go back to attacking the body and bring the disease back. Even the bone marrow transplant from Kira isn't a good enough treatment because Cosima's symptoms have progressed to the point that such a transplant only slows down the process rather than stops it.

Gene therapy (inserting a genetic sequence as a treatment, much like a drug, to correct the gene causing the illness) is really the best option because it is the most long-term solution, with the potential to fully reverse the effects of the disease and prevent it from coming back.

CASE STUDY:
M.K. (VEERA SUOMINEN)

ID: 3MK29A
DOB: March 3, 1984
Birthplace: Helsinki, Finland
Status: Deceased

BETH

There. It's a nice face.

(4.01 "The Collapse of Nature")

Veera Suominen makes her first appearance in issue #5 ("Rachel") of the *Orphan Black* comics as a patient of Dr. Nealon. As a child, Veera was trapped in the same fire that destroyed the Duncans' experiments at their lab in Cambridge, England. She survived but sustained damage to her lungs and severe burns to the right side of her body, including the right side of her face. Along with Katja Obinger, she's one of the sole clones to have survived the "Helsinki incident" — which took place when the clones were teenagers, and which Rachel alludes to for the first time in "The Weight of this Combination." When at least six self-aware European clones threatened to go public, Dyad and Topside had the clones assassinated to protect their experiment.

When Veera later appears in season four of the show, she goes by a new name — M.K. — and we learn that she has been in hiding since the Helsinki incident and takes great pains to remain hidden, from donning a novelty sheep mask to keeping all communications limited to three minutes. She's a talented hacker who tried to help Beth in her Neolution investigation, but when Sarah finds her, she's hesitant to come to the other clones' aid. Her burns have long since healed, leaving distinctive scars that she often hides by parting her hair so that it falls over the right side of her face. Without knowing Veera's history with the Duncans' lab explosion, it would be easy to assume that her scars were from Helsinki.

Skin is the body's first line of defense against the outside world: it's a physical barrier against bacteria and other microbes, and it hosts "good" microbes, often referred to as flora or microbiota, which also serve to protect the body against unwanted pathogens. Skin microbiota protects us by outcompeting harmful microbes for nutrients (so that they can't thrive on the skin), by producing chemicals against them, or by triggering the immune system against them. But while most of them are harmless or helpful, some of the microbes are opportunistic and will become harmful if given the chance. Major skin burns are dangerous because they are a breach of the physical barrier that protects the body and because they are prone to infections. The microbiota that usually live innocuously on the skin are suddenly a threat that the body is less prepared to fight off. These infections can happen very quickly and can snowball into septic shock and organ failure. Another potential complication of major burns is fluid loss through the damaged skin, which can lead to kidney failure. Basically, burns introduce all kinds of bad.

Skin grafts to treat burns — which Veera receives under the care of Dr. Nealon in the comic-book issue "Rachel" — are critical, then, because they can be used to reduce the concentration of bacteria in the burn wounds and to help the body retain fluids.

NEALON

She was badly burned in the fire.
Luckily, we had an ample source for
skin grafts.

(Orphan Black #5 "Rachel")

Newer technologies allow us the possibility of taking skin cells from the burn patient to grow new skin cells in sheets in a lab, or even the use of synthetic skins, but back when Veera was burned, her skin graft would have been harvested from a person using a tool called a dermatome, which can remove a layer of skin. The thickness of the skin depends on the needs of the person receiving the graft. You might have heard skin burn severity described in terms of *degrees*; it's more

accurate to think in terms of *depth*. First-degree burns are superficial and only affect the topmost layer of skin, called the epidermis. The burned skin might peel after a day or so, but it won't blister. Second-degree burns, also known as partial-thickness burns, do blister, and the damage affects the epidermis and part of the dermis, the layer below. Third-degree burns, or full-thickness burns, extend through all layers of skin. If a burn is referred to as a fourth-degree burn, it means that other tissues that lie below the skin (like muscles, tendons, or bones) have also been damaged. Skin grafts can be used to treat deep partial- and full-thickness burns because they take longer to heal and have a much higher risk of developing infection than superficial burns. Accordingly, skin grafts can be either partial (split-thickness) or full-thickness. Split-thickness grafts are more common because they can treat larger areas and involve less tissue (meaning less requirement for vascularization, or blood-vessel formation, in the graft as the burn heals). Full-thickness grafts are useful in smaller, highly visible areas with good access to vascularization, like the face, because they contract less as they heal and maintain characteristics closer to "normal" skin.

As part of skin graft preparation, the skin sample may be passed through a mesher. A mesher does exactly what you'd expect: it punches holes into the skin, so it can stretch like a mesh and cover a larger burn area. Preferably a skin graft is taken from the burn victim's own body, usually from an area with a large enough surface area, like the thigh or buttocks (this is known as an autologous transplant or autograft); skin can be harvested from the same donor site more than once, as new layers of skin grow (although there is a limit to how many times you can do this). Using skin that isn't foreign avoids the risk of the grafted skin being attacked by the immune system.

Since Veera's burns were severe and covered a large part of her

body, her skin grafts had to be sourced from another person (known as an isogeneic transplant if the donor is genetically identical, or an allogeneic transplant, or allograft, if the donor is not). Autografts are meant to be permanent skin replacements while allografts act more as a temporary skin to close a wound, a protective barrier preventing loss of fluids while the burned skin heals. Allografts are eventually rejected by the body's immune system. Allogeneic transplants can come from known donors or skin banks (like a blood bank . . . but for skin: skin banks receive grafts harvested from the cadavers of people who agreed to be organ donors, test them for communicable diseases, and preserve them until they are needed). Veera's grafts were likely isogeneic: based on what Nealon tells Rachel, it sounds like the grafts were taken from another Leda clone.

CASE STUDY:
KRYSTAL GODERITCH

ID: 843V90
DOB: May 5, 1984
Birthplace: Sudbury, Ontario, Canada
Status: Alive

KRYSTAL
Yeah, and I'm being personally targeted because I'm a whistle-blower, right? I'm like that guy who moved to Russia.

(4.05 "Human Raw Material")

Krystal Goderitch is a naive clone despite being in the thick of Neolution conspiracy. Even when she is officially inducted into Clone Club by Sarah in "From Dancing Mice to Psychopaths," she seems to be willfully ignorant of the fact that she is a clone. She's been labeled a non-threat by Neolution and the Leda clones alike, but within only a few weeks, Krystal uncovered nearly as much information as Sarah and the other clones put together over the course of six months. Krystal prides herself in particular on her ability to dig up and expose Neolution's unethical and illegal activities, though she may not get all of the facts right in terms of how they connect to clones and to herself. Everyone from Felix to Susan dismisses Krystal for her naiveté, but she makes some salient points.

In "Human Raw Material," Krystal has somehow connected Brightborn to Dyad and Neolution and is doing a little independent sleuthing at the same time that Cosima and Donnie have gone undercover to learn more about Brightborn's seemingly all too perfect offerings. At first, it seems that Krystal has everything figured out, but then she waves a tabloid clipping in Donnie's face, claiming a link between Neolution and stem cell cosmetic treatments that caused a woman to

grow teeth on her eyelids. She's dismissed again as knowing nothing (but not before she nabs a few cosmetics samples and tucks them away in her purse). The thing is: this particular claim is based on a real incident.

In 2009, a woman in L.A. underwent a non–FDA approved cosmetic procedure that basically extracted her adult stem cells (mesenchymal stem cells) via liposuction and re-injected these cells into her face and around her eyes as a sort of stem cell facelift. Mesenchymal stem cells can develop into fat, cartilage, and bone, among other cell types, and the goal of this treatment was to have the stem cells differentiate into brand new tissue. But instead, the stem cells differentiated into bone cells and proliferated into tiny bone shards that made blinking difficult and painful, grating against each other to cause a clicking sound "like a castanet" whenever her eye was opened (much like what Krystal describes to Donnie).

During the woman's procedure, the surgeons had also injected dermal filler, a chemical that is commonly (and safely!) used in cosmetic procedures to reduce the appearance of wrinkles. One of its main components is calcium hydroxylapatite, a mineral that is found in human bones and teeth. Hydroxylapatite has been used experimentally to treat damaged bone tissue and to coat implants such as hip replacements or dental implants. It is thought that coating these implants assists with osseointegration, or having bone grow right up against the artificial implant and interact with it without any soft tissue interfering in between. All of this sounds like it has no business in a cosmetic facial treatment. So, what's hydroxylapatite doing in facial dermal fillers like Radiesse? When it's injected into the dermis, calcium hydroxylapatite is meant to act as a mineral scaffolding upon which collagen can grow to fill in and soften lines and wrinkles; the mineral eventually breaks down into calcium and phosphate components that the body can metabolize. In the case of the stem cell eye treatment, the hydroxylapatite in the dermal filler was similar enough to bone tissue to trigger the injected stems cells to form bone cells instead of the desired tissue. In this case, the bone shards were extracted, but in some cases living stem cells may linger on and develop unexpectedly into bone (or other) tissue again.

Many cosmetic products such as lotions and other topical skin treatments proclaim their use of plant stem cells as a rejuvenating ingredient. Cosmetics connoisseur that Krystal is, we're sure she would tell you that these claims are totally bogus. For one thing, plant stem cells, whether from Swiss apple, or rose, or Gamay grape, or any other plant, would not be especially helpful for activating cell growth in your skin by virtue of them being stem cells from *plants*. As far as has been measured, plant cells and human cells just don't communicate. Human cells don't possess receptors to receive the chemical signals that plant cells produce. Besides, most products purported to contain stem cells do not contain living stem cells, but rather stem cell extracts. It's just too difficult to sustain living cells in a lotion.

There are companies out there that offer lotions and eye creams derived from your own stem cells. One of these is Personal Cell Sciences, based in New Jersey, which offers a line of adult stem cell products called U Autologous. According to their website, they collect your stem cells via liposuction, much like in the procedure that led to eyelid bone growth. Some of these cells are cryogenically frozen and stored for future use, while others are cultured and grown before being blended into skin care products.

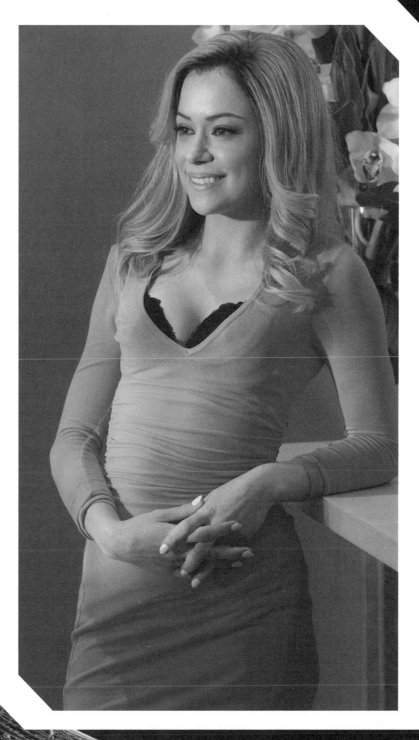

The FDA has yet to approve any stem cell cosmetic products like these, and the science just doesn't support their advertised effectiveness. The thing is, many stem cell cosmetic products sit in a sort of legal gray area: unlike drugs, cosmetics do not require FDA approval before hitting the market and appearing on shelves in your local pharmacy. But if human cells or tissue were incorporated into the active ingredients, especially if they were intended to be living stem cells, then the product would technically classify as both a cosmetic and a drug. The FDA has stepped in more than once in recent years to issue warnings to manufacturers distributing stem cell cosmetic products.

Now, if you applied human stem cells directly onto your skin, you probably wouldn't get whatever effect you desired; you'd probably just be left with a layer of dead stem cells. When it comes to cosmetics that claim to encourage new cell growth, it's not exactly the cells themselves that do the trick; it's the by-products that they produce, such as cytokines and growth factors. Cytokines are a family of proteins that allow for cells to communicate with each other, and they have myriad roles in the body, from stimulating cell division, to cell metabolism, to immune function. And unlike plant by-products, they can interact with receptors on skin and other tissue cells. When it comes to using cytokines in a skin product, right now there's no guarantee that the cytokines involved are specific to cell repair — or if they are instead associated with tumor growth. While rubbing cyto-kines and growth factor directly onto your skin may be less risky than injecting it into your skin, it's unregulated and not without its risks.

As it turns out, Krystal wasn't wrong in suspecting a tie between Dyad and sketchy cosmetics. In season five, after her roommate has an extreme reaction to a stolen cream, she uncovers that Dyad has partnered with BluZone cosmetics to produce a line of experimental treatments with a dermal delivery system — essentially gene therapy that can be absorbed through the skin. Len, the CEO of BluZone, describes the treatment as the next big thing in regenerative therapy (taking the usual cosmetic claim of "younger-looking skin" more than a step forward). Mrs. S, however, later suggests that these Dyad-brand cosmetics might also be used to deliver the Castor pathogen to specific populations. This double-whammy feeds right into Neolution's goal to drive human evolution and control which "desirable" traits are spread through a population (and which traits are culled).

Chapter 7

"AND WE, HERE, SHALL DRINK FROM THE FOUNTAIN *FIRST*"

PROLONGEVITY AND REGENERATION

> **RACHEL**
> [...] the so-called "Fountain of
> Youth" gene. We're decades ahead of
> popular science. You're both lucky
> to be here; it's a momentous time.
>
> *(5.05 "Ease for Idle Millionaires")*

In the season four finale, Cosima thinks she's dying. Between the fact that she was found sick and hypothermic in the woods and that she finds out Delphine is not dead after all, she's not in any sort of shape to notice what's going on in the strange village she's been brought into.

By the season five premiere, she's had a chance to rest up, and Cosima, being Cosima, gets curious and decides to poke around the village (which she learns from one of its inhabitants, a woman called Mud, is named Revival).

Revival presents a whole new facet of Neolution. The village is largely self-sufficient: living off the grid, generating their own power, raising livestock and fish and growing their own food in green-houses. The people living there aren't Freaky Leekies with bleached hair and white contact lenses, and they aren't rich people angling for designer babies or functional tails. Mostly they're people who have made personal pilgrimages to seek new lives, cures for their diseases, and the promise of Westmorland's rumored "fountain of youth."

DELPHINE
This entire island is a decades-long prolongevity study.

COSIMA
And everyone here is participating.

(5.01 "The Few Who Dare")

The way that the Revival villagers see it, Westmorland is a savior figure; they grant him a cult following, calling him the Founder and viewing themselves as His children.

As the Father of Neolution, of course Westmorland is obsessed with participatory evolution, and many of the people who are referred to Revival have unusual biology, such as rare cancers, that he's keen to analyze as a means of uncovering mutations that could serve his purpose. It's clear that his eye is on the prize of immortality — or at least an extremely prolonged lifespan. This isn't an uncommon goal for futurists and transhumanists or, really, for the average person: who wouldn't want to at least try to live for as long as possible? As Delphine tells Cosima: if you want to genetically "improve" the human race, life extension is the first principle. It's the *heart* of Neolution. The problem with craving prolongevity, another term for life extension, is that aging isn't even close to a cut-and-dry process.

Cosima and Scott's initial research turns up some interesting material. While exploring Westmorland's cabinet of curiosities in "Clutch of Greed," Cosima comes across a photo of him just chilling with Sir Arthur Conan Doyle in 1894. (Doyle was studying medicine in the late 1800s and publishing his Sherlock Holmes mysteries by the 1890s.) The photo also features a Galapagos giant tortoise, known to be one of the longest-living vertebrate animals, with lifespans stretching long past one hundred years in the wild. Westmorland tells Cosima that Arthur Conan Doyle rode one of Darwin's tortoises right after that photo was taken. It's a funny story, but the truth is Darwin himself reported doing the very same thing. Galapagos tortoises were among the animals that Darwin took note of during his voyages on the HMS *Beagle* in 1835. It was on the Galapagos islands that Darwin famously developed his theory of evolution, and even though finches played the largest role in this development, he also noted the differences that he saw in the tortoises on different islands. That said, he probably had no idea about their exceptional lifespans and definitely didn't realize their importance to his research at first. He brought some 30 tortoises onto the *Beagle*, not to study but as fresh meat for him and his crew to eat on their travels, after learning that tortoise meat was a staple to island inhabitants. (Darwin wrote that they also drank fluids from the tortoises' bladders and pericardia when freshwater was scarce.) And when Darwin was observing the tortoises on land, he was amused by how easy it was to sneak up on them and startle them. He would climb onto the tortoises' backs, knock on their shells to get them moving, and yep: *ride them around for fun* (but probably not very far — Darwin also admitted that keeping balance was a challenge).

Not all of Darwin's tortoises ended up eaten. Arguably the most famous of them, Harriet, is believed to have been collected by Darwin in 1835, when she would have been about the size of a dinner plate. She spent a cold and cloudy bout in England, before John Wickham, a pal of Darwin, brought her to Australia in 1842 to enjoy a more Galapagos tortoise–friendly climate. Other historians are dubious of this and claim that it's more likely that Harriet was brought to Australia sometime later by a whaler.

Listening to Westmorland's story, Cosima remarks that she thought that all of Darwin's tortoises had died. Indeed, Harriet died in 2006, under the care of Australia Zoo. She died at the ripe old (estimated) age of 175. Westmorland refers to the tortoise in the photo as Beatrice . . . which isn't one of Darwin's named tortoises. Actually, Harriet wasn't even named by Darwin — she was called "Harry" in the early 1900s after Harry Oakman, the groundskeeper at the Brisbane Botanical Gardens, which she called home at that time, and was renamed later. Up to two other Darwin-adopted tortoises are rumored to have existed (and were called Tom and Dick — *get it*?), but their histories are even fuzzier than Harriet's. Westmorland's Beatrice might be a renamed Tom or Dick, or she's a completely phony story that he's feeding to Cosima, couched in just enough known Darwin lore to seem plausible.

Cosima and Scott also found research papers authored by Westmorland from as far back as 1867, not to mention the paperweight *On the Science of Neolution* that we first see in Susan Duncan's island house, which was published in 1872. And then, of course, we have the man himself, living on a mysterious island in 2013.

All signs seem to point to a 170-year-old man. Could that be possible?

When P.T. Westmorland was born in 1842, the average life expectancy in England and Wales was hovering around the 40-year mark. As of 2015, global life expectancy was 71.4 years, and 29 countries pointed toward average life spans of 80 years or more. It's worth mentioning that women typically outlive men by a significant margin, a fact that probably stirs Westmorland's rage — not only is immortality beyond his reach, but by virtue of his own biology he's already at a disadvantage. Globally, life expectancy has been increasing by about three years per decade since the 1950s (except for the 1990s, which saw the rising HIV epidemic in Africa and the collapse of the Soviet Union). Even just between 2000 and 2015, we've seen a global jump in life expectancy of five years. This is mostly thanks to improvements in public health worldwide, including sanitation, antibiotics, and improved treatments for conditions like high blood pressure. Longevity is heading in the direction that Westmorland craves; it's just not heading there fast enough for his taste.

By convention, centenarians are people who live for 100 to 110 years, and anyone who lives beyond that 110-year mark is called a supercentenarian. As you might guess, centenarians (and especially supercentenarians) aren't super common. There are some populations that more frequently report centenarians and supercentenarians among their ranks. Demographers Gianni Pes and Michel Poulain identified five geographical areas that seem to have the highest concentrations of people living into extreme old age: Sardinia, Italy; Okinawa, Japan; Nicoya, Costa Rica; Icaria, Greece; and Loma Linda, California, USA (specifically the Seventh-day Adventists in this last case). They called these longevity hotspots "Blue Zones" (which is absolutely the inspiration for the BluZone cosmetics company that partners with Dyad to produce tissue regeneration in a lotion bottle).

Pes and Poulain observed the groups of people living in their Blue Zones and suggested common lifestyle factors that might be contributing to their longer lifespans. Boiled down, none of these factors is especially surprising: these groups tend to practice regular physical activity that is part of their day-to-day living instead of *exercise*; they smoke less; they practice moderate caloric intake, favor plant-based diets, drink moderate amounts of wine; and they are socially engaged with their families and communities. All the sort of stuff that we already *know* promotes a healthy lifestyle. Researchers have also taken to studying the genomes of these populations (and those of other long-living populations outside of the Blue Zones, like the Ashkenazim) to determine any genetic component to their lifespans and better health into late life. On average — inherited and infectious diseases aside — humans tend to enjoy good health until about age 45 or 50 before age-related conditions start cropping up. The oldest of the old seem to have good health for much longer, and although a healthy lifestyle could certainly contribute to that, when you consider that some of these Blue Zones represent somewhat isolated communities, it's likely that genetics also plays a role.

ON AVERAGE — INHERITED AND INFECTIOUS DISEASES ASIDE — HUMANS TEND TO ENJOY GOOD HEALTH UNTIL ABOUT AGE 45 OR 50 BEFORE AGE-RELATED CONDITIONS START CROPPING UP.

In season five, Delphine returns from a trip to Sardinia and complains about collecting stool samples from centenarians. She calls the Sardinian gene bank that Westmorland purchased a goldmine, and for his purposes, it absolutely would be.

Sardinia was the first of the five Blue Zones that Pes and Poulain identified and is, of course, known for its population of centenarians, especially its high incidence of male centenarians. Its biobank is likewise internationally known as one of the oldest and largest gene repositories in the world. The bank has been collecting biological samples from its citizens since the year 2000, as part of a government-funded project to identify genes that might confer protection against aging. In September 2016, the project made headlines when it was thought that thousands of DNA samples had been stolen from their facilities. Luckily, it was a false alarm: the samples were found at a Sardinian hospital, where scientists had moved them some years earlier. The "theft" proved a segue into a larger discussion about the ownership of this genetic information when it came to light that a for-profit London-based biotech company, Tiziana Life Sciences, acquired the biobank in 2016 and established LonGevia Genomics in Sardinia to oversee it. In a statement about the acquisition, Tiziana's CEO and founder Gabriele Cerrone said that sequencing the unique collection of samples from Sardinian citizens provided an opportunity "to generate valuable insights into gene regulatory networks, genotype-phenotype linkage, and gene-environment interactions that will feed into and inform our drug discovery and diagnostic programmes." In short: they're a biotech company. They're in the business of designing biotech treatments and tools. They plan to use the genetic info that they've bought to create designer drugs and diagnostic kits that they can then sell for profit (profit that the people who donated the cells will never see).

COSIMA

All this suffering just so you can extend your life. You're not even a hundred and seventy. That's all smoke.

(5.05 "Ease for Idle Millionaires")

If P.T. Westmorland's claim of being 170 seemed farfetched to you, you're not alone. Westmorland *was* a real person (at least, in the *Orphan Black* universe), but he disappeared around 1894 and was declared dead in 1898. When he resurfaced in the 1950s, he wasn't the same man at all. This isn't a metaphor: the new P.T. Westmorland was patently *not the same dude*. Rather, Westmorland 2.0 was an ambitious Cambridge student, John Mathieson (although, as Susan is quick to point out, academically he was a "distinctly middling" student). Mathieson hoped to take on Westmorland's ethos and resurrect the Father of Neolution's ideals for a new wave of transhumanist research. It was at university that he teamed up with Susan Duncan after recognizing her brilliance in her field (although she was Susan Wakefield at the time) and then eventually with Ethan Duncan and Virginia Coady, who would all become instrumental in his plan to take human evolution into his own hands.

This brings up another issue: at the point where Westmorland (we'll keep calling him by his chosen name to keep things from getting confusing) is trying to master his own mortality, he's already, well, *old*. He may not be the supercentenarian that he claims to be, but he's at a point where age-related illnesses are following him like a shadow. Cosima calls him out on this when she discovers the extent of his experiments on Revival villagers, and he doesn't dispute her point. In "To Right the Wrongs of Many," he appears to have a stroke, though he takes such offense that he kills the doctor who makes this diagnosis (well, also because that doctor dared imply that he was old). We see Westmorland become increasingly desperate to seek out and isolate a treatment for aging, but at this point in his life, it's not a benefit to PTW anymore to simply isolate genes that will grant him a longer life — not if it means that these extra years will be plagued with illness. It makes much more sense for him to try to find a way to rewind the clock.

COSIMA
You're dying, P.T. This is your
legacy.

(5.05 "Ease for Idle Millionaires")

KEEPING YOUR CELLS ALIVE

A key to living a long life is having a body that's up to the task of maintaining itself and properly functioning upward of 90 years. That means all specific cell types and tissues must perform their proper functions, the genome must be properly replicated without mutation or error every time a cell divides, and any problems that arise — such as cell death or overgrowth — must be dealt with.

Normally, when a cell detects some sort of dysfunction, the cell undergoes a process known as apoptosis: programmed cell death. The cell kills itself in order to overcome the dysfunction, sacrificing itself to save the whole, and other nearby cells replicate or differentiate to take the dead cell's place. If no nearby cells are able to take up the slack of the dead cell, however, this results in the complete loss of the cell and minor degradation of the tissue.

When a cell exits the cell cycle — it no longer undergoes division and replication and can therefore no longer prevent the degradation of its tissue due to apoptosis — the cell is called senescent. Some cell types, such as neurons, become senescent during development and never divide during a person's life. Other cell types — muscles, skin, fat, et cetera — become senescent over time, causing the tissue to undergo degradation with age. Part of the reason for this attained senescence is a phenomenon known as the Hayflick limit: the natural limit of rounds of division a cell will undergo before exiting the cell cycle. The limit is thought to be set by telomere length and its subsequent shortening (as we discussed in terms of Charlotte and the clone disease in chapter six). Telomere length determines the number of divisions a cell can undergo before damaging its genome.

In order to prolong someone's lifespan significantly, it would be necessary to overcome senescence — reverting senescent cells back into the cell cycle, where they can continue to replicate and maintain tissue integrity.

PARABIOSIS

IRA
Young blood, for old veins.

SUSAN
He'll try anything to extend his
pitiful life.

(5.06 *"Manacled Slim Wrists"*)

Another method used to reverse the aging of cells is known as parabiosis, which literally means the joining of two individuals anatomically. In an anti-aging study, researchers joined a young mouse and an old mouse by sewing their skin together, allowing blood to flow freely between both animals. While the young mouse showed no obvious changes, the old mouse showed improvements in healing and cognitive function. This supports the concept of neoteny — that is, when juvenile traits are preserved into later stages, degeneration due to senescence can slow down.

Westmorland takes this research to the next level by regularly injecting himself with young blood. This modified version of parabiosis allows the cells within PTW's aging tissues to be exposed to elements within the young blood, like proteins, healthier immune cells, maybe even young stem or progenitor cells, any of which might give his cells the opportunity to essentially reverse some of the aging process. While this may seem pretty ingenious, it is not the most efficient method of increasing lifespan. Westmorland requires frequent sessions of parabiosis (much like a patient in kidney failure must undergo dialysis), and the process requires him to be hooked up to a machine for long stretches of time. Parabiosis in this manner also requires a continuous source of young blood, which would rely heavily on a large number of willing and able blood donors (which might not be the easiest thing to obtain on a remote island). Therefore, for PTW, parabiosis is simply a temporary pause on his aging process while he pursues more promising avenues of research.

LIN28A: NEOLUTION'S "FOUNTAIN OF YOUTH" GENE

COSIMA
Lin28a. Tell Westmorland I found his fountain.

(5.05 "Ease for Idle Millionaires")

PTW's "fountain," or what he believes is the key to live forever, is in the manipulation of the gene Lin28a. This gene (an actual gene!) encodes a protein that regulates developmental timing and stem cell self-renewal. That is, Lin28a is important in determining when certain processes stop or start, as well as controlling the size of the stem cell pool during development. If these processes aren't carried out properly, severe developmental disorders can arise due to the malformation of various tissues.

Lin28a is also one of a handful of genes used in the process of reprogramming differentiated cells back to pluripotent stem cells.

Let's say you take a biopsy of skin cells from a person and grow these cells in the presence of reprogramming factors (known as Yamanaka factors after the scientist who first discovered this process). The cells will become stem cells that have regained the ability to differentiate into a multitude of cell types. Attempts have been made by researchers to overcome senescence with this process. One group of researchers introduced Yamanaka factors into aging mice (specifically mice that had a genetic condition called progeria, which ages the body prematurely). These factors bind to regions of DNA (called promoters) that initiate transcription, the first step of gene expression. When the mice received a pulse of these factors, their reprogramming functions took effect and the mice lost signatures of aging in their cells, their lifespans increased, and their tissues showed improved regeneration.

Through studying the reprogramming ability of Lin28a, researchers also discovered that this gene has a role in healing and regeneration. Mice with a modified version of Lin28a were able to regenerate tissues, such as their ears and toes when they were clipped. This enhanced ability is thanks to Lin28a's role in controlling cell metabolism, increasing the production of cellular energy enough to meet and exceed the needs of wound healing and regeneration. Normally, Lin28a is expressed during embryonic development, but is shut off shortly after birth, which explains the naturally enhanced ability of younger organisms to heal compared to their older counterparts. When this gene is altered to remain on long past birth, these increased healing abilities are maintained.

Helena's unborn babies (the future Arthur and Donald) are a prime example of the benefits to be had from a dysregulation of Lin28a. In the season five premiere, Helena is impaled by a sharpened stick after saving Donnie from Neolution agents. The stick pierces her abdomen. Donnie brings Helena to the hospital where she has emergency surgery to remove the stick, which has punctured the amniotic sac as well as the chest of one of her fetuses. To the medical staff's surprise, neither of the fetuses seem to be in distress — their heartbeats and amniotic levels are good — despite the injury. A second ultrasound is ordered and Helena's nurse finds that what seemed too good to be true is, in fact, true: Helena's son has completely healed himself, showing no signs of having recently been impaled by a stick.

The twins benefit from accelerated healing because of adjustments made to Lin28a, which they inherited from Helena (more on this below). Although the twins were still in utero at the time of injury, when Lin28a is normally expressed, their genome possesses changes to the gene that improved upon its normal function, allowing the fetus to fully heal from a serious wound in a matter of hours.

NEOLUTION'S MONSTER

Introduced in the season five premiere as the mysterious beast that attacks Sarah in the woods, and feared but largely dismissed by the

Revival villagers as a bear, we learn the true story of this "monster" in "Ease for Idle Millionaires." He was once a young Latvian orphan named Yanis. Before Project Leda was conceived, Westmorland became interested in Yanis's biology due to the boy's unusual healing ability, caused by a mutation in Lin28a. Yanis provided him with his first subject upon whom to carry out studies of this potentially miraculous gene. While Helena's twins reap the benefits that can be unlocked by manipulating the human genome, Yanis demonstrates how terribly wrong everything can go when a scientist decides to play god.

In order to understand how to harness the "miracle" of Lin28a, PTW performed experiments on Yanis, manipulating and mutating Lin28a in various ways, often resulting in horrible outcomes, such as his tumors. Scientifically, Yanis was invaluable to unlocking this "fountain" and providing Westmorland with answers. But Yanis was treated no better than a glorified lab rat, and his fate was as inconsequential to PTW as that of the Helsinki clones to Rachel. To Westmorland, Yanis was simply a means to end, the key needed to unlock his scientific legacy. Yanis's transformation into a "monster" is a classic example of a gene whose regulation has gone awry: uncontrollable growths, cognitive changes and deficits, and a plethora of medical issues beyond treatment. And Yanis's years of scientific torture illustrate just how necessary are the rules, codes, protocols, and laws in place today regarding research involving human subjects or human samples. His storyline is a manifestation of many of the themes that run through all five seasons of *Orphan Black* — those of autonomy, the boundaries of ethical science, and the vital refrain of "my biology, my decision."

HENRIETTA LACKS

Henrietta Lacks is probably the most famous real-life example of the many ethical issues faced when research involves human samples. Henrietta Lacks was a poor black woman suffering from cervical cancer; she eventually succumbed to the disease in 1951. While she was in treatment at Johns Hopkins Hospital, unbeknownst to her, her cancer cells were biopsied and cultured, giving rise to the HeLa cell line — the first immortal cell line and arguably the most important cell line ever established for medical research.

HeLa cells have since and are still used across the globe by scientists, as they represent one of the most well-characterized and widely studied human cell lines. As a result of studies involving HeLa cells, scientists have been able to make huge breakthroughs in biology, especially disease treatments. For example, it was with HeLa cells that Jonas Salk first developed the polio vaccine. Advances in treatments for cancer, HIV/AIDS, and radiation poisoning, as well as study of gene mapping and protein dynamics, were all made using HeLa cells.

Despite all the scientific good that has come from the use of this cell line, the cells were taken from Henrietta without her consent, and the use of her cells in research was unknown to her family until the mid 1970s. Henrietta's family, descended from slaves of the tobacco farms in the South, never saw any compensation for the use of Henrietta's cells or for the plethora of advancements and commercial uses that came from studies involving HeLa cells.

A Supreme Court decision has since determined that discarded tissue and cells are no longer that person's property, and although Henrietta and her family did not give permission for the cells to be taken, permission at the time was not a standard requirement before harvesting biopsies. The family still has not received any financial compensation for the use of HeLa cells, although they did obtain the right to decide who has access to the HeLa genetic sequence. Henrietta Lacks is still the focus of many discussions today about consent and privacy when it comes to scientific research, shining light on the many gray areas that are so often overlooked.

AISHA

Even without Westmorland's mad science meddling, naturally occurring mutations and dysregulation of Lin28a can give rise to a number of cancers, a problem not uncommon with genes involved in tissue and cell growth and division. When expression of Lin28a is misregulated, the result can be an overgrowth of cells into a tumor.

In the season five premiere, we meet Aisha, an Afghan girl with cancer who has just arrived at Revival in hopes of a cure. Aisha has a pediatric cancer of the kidney, called Wilm's tumor, resulting from misexpression of Lin28a. While Aisha and her mother believe she is at Revival to be cured, most likely Aisha was brought to the island because of PTW's interest in Lin28a, providing him with a chance to study his "fountain" gene further before she succumbed to her disease. PTW could learn from Aisha's mutation, better understanding the ways Lin28a can be manipulated and what changes should be avoided. After all, PTW is looking to live forever; he needs to make sure he doesn't accidentally give himself terminal cancer.

SO, WHERE DOES KIRA FIT IN?

From almost the beginning of the series, we knew that there was something . . . special about Kira. After being hit by a car in "Entangled Bank," she's found to have little injury beyond some scrapes and bruises in "Unconscious Selection": no broken bones, no damage to her internal organs. Felix dismisses the miraculous lack of injuries and quick recovery, citing that young children are "made of rubber," while Cosima refers instead to the amazing regenerative abilities found in lizards — a concept with which Sarah refuses to engage.

Her daughter is *not* a lizard.

While we agree — Kira is *definitely* not a lizard — her biology was actually inspired by another animal with amazing abilities. You've probably heard that some lizards and salamanders can drop a part of their body, like a tail, to escape a predator and then regenerate a brand new one. How does that work? Let's look at the salamander, which is probably the quintessential regenerating vertebrate. Within hours of the salamander's tail being removed, cells gather at the site to generate a blastema. A blastema is a mass of cells with the ability to (re)generate tissue. If the term seems familiar, you might be thinking of our earlier discussion of blastocysts, the balls of cells that eventually develop into embryos. If you guessed that the root term "blast" might have something to do with tissue growth and development, then you'd be right on target. The term originates from the Ancient Greek *blastós*, meaning "germ" or "sprout."

Blastemas might sound like the same thing as stem cells, but unlike stem cells, blastemas are not exactly undifferentiated. They're more like *de*differentiated, and unlike pluripotent stem cells, which have the potential to become almost any tissue, the blastema is formed by a collection of restricted progenitor cells. These cells have a sort of "memory" of their tissue of origin. As the blastema forms, genes responsible for the body plan (namely, Hox genes) become activated. Hox genes form a group of genes responsible for directing the body plan in a developing embryo, and when they're activated in the case of regeneration, their job is the same: to guide where a specific body part should develop (i.e., replace a missing tail with a new tail and not, say, a foot). Experiments with Hox gene mutations in fruit flies famously produce body parts where they shouldn't be, like fly legs in place of mouth parts. Interestingly, evolutionary biologists believe that the loss of specific Hox genes (those encoding for forelimbs and hindlimbs) explains the loss of limbs in snakes. So, the collection of cells in the blastema are directed by these genes to form a replacement tail, and new muscle tissue, nerve tissue, and blood vessels develop with it and reconnect to existing muscle tissue, nerve tissue, and blood vessels at the site of injury. And voilà! a perfect, functional, regrown replacement.

Although many animals, such as reptiles, amphibians, worms, and echinoderms (like starfish), have shown the ability to regenerate tissues and even whole limbs, most mammals just don't have that ability. If a mammal receives major tissue damage, say, to its skin, it can heal, but that healed tissue will become scar tissue and won't have all of the functions of normal skin, like hair follicles and sweat and sebaceous (oil) glands. But there is at least one exception to the rule.

Enter the *Acomys* mouse, also known as the African spiny mouse. This is the pet that Rachel gifts Kira in "Beneath Her Heart,"

which she uses to explain Kira's regenerative powers. These mice have skin that tears really easily when handled — skin reportedly 20 times weaker than the skin of a typical lab mouse — and it probably evolved this way so that the spiny mouse could quickly shed its skin to escape predators, the same reason why some lizards drop their tails. At least two species of spiny mice have been shown to repair this damaged skin and regenerate hair follicles and glands — *fresh skin*, not scar tissue — and can even regenerate fat cells and cartilage. These regenerative powers are incredible, but they do have their limits: spiny mice can't regenerate muscle cells.

So if at least one type of mammal shows regenerative power, maybe there's a chance that humans can harness it, right? That's what Westmorland was banking on when he and his team started tinkering with the Project Leda genome.

In season five, we learn that one of the synthetic sequences introduced into the Leda clones' "perfect human baseline" genomes was intended to produce heightened tissue regeneration in the form of accelerated healing. Immortality-obsessed Westmorland hoped that this sequence would be expressed in the Leda clones, so that he could then harvest and apply the mutation to himself. While the offspring of the Leda clones represent the ability to harness the function of Lin28a to improve on the human genome, it wasn't simple work for the Project Leda scientists to just pop the mutation into the Leda clones' genome. Understanding the function and regulation of Lin28a required years of experimentation and manipulation.

The Ledas' resulting lack of healing ability after so much effort proved to be a disappointment (how he tested for this in the Leda clones is unclear, but we'd rather not think about it). Even though the Leda clones can take quite a beating, survive, and heal — an understatement when you consider that over the course of five seasons, Helena got stabbed by a piece of rebar in the liver, shot point-blank in the chest, and impaled by a sharpened stick; Sarah got stabbed, not to mention beaten down more than once by fist, boot, and walking cane; and Rachel got a pencil shot into her eye — they don't seem to show any sort of superhuman healing ability. And as Helena's self-inflicted wings show, when first-gen Leda skin heals, it scars.

Project Leda clones were never intended to produce children, but by Sarah's and Helena's chance fertility, we get to see that the sequence did successfully code for accelerated healing — it just needed a second generation to manifest. We see this in Kira, and we see this in Helena's babies.

The fact that the healing trait does appear in the clones' unplanned daughter generation is clearly of special interest to Neolution. Also important is the fact that their Lin28a mutation seems to be self-regulating: it expresses the tissue regeneration trait but it keeps itself in check, in contrast to the uncontrolled cell growth and division we see in the version of the mutation that has caused cancers in Revival villagers like Yanis and Aisha.

> BY SARAH'S AND HELENA'S CHANCE FERTILITY, WE GET TO SEE THAT THE SEQUENCE DID SUCCESSFULLY CODE FOR ACCELERATED HEALING — IT JUST NEEDED A SECOND GENERATION TO MANIFEST.

RACHEL

If we want to know if our lab rats'
tails will grow back, we damn well
will cut them off and see!

(4.10 "From Dancing Mice to Psychopaths")

Rachel makes it clear to the Board in "From Dancing Mice to Psychopaths" that she's intent on moving forward with Neolution's experiments with the clones' biology. Cosima has developed a cell line as a source for a Leda disease cure, but Rachel sees an opportunity to use the cells for reproductive cloning: create a new generation of human clones free of the disease genotype and produce a new generation of test subjects prime for testing and manipulating for desirable mutations. At this point, it sounds like she's planning on putting Evie Cho's maggot-bots into these hypothetical clones.

But where the Castor and Leda clones fall short in desirable traits, Kira Manning and Helena's babies offer so much new potential. In "Gag or Throttle," Rachel shows off Kira to the board members and describes her plans to biopsy Kira's liver, lungs, and stomach and study her cells. The liver, lungs, and stomach — specifically the stomach mucosa, the tissue that forms the inner lining

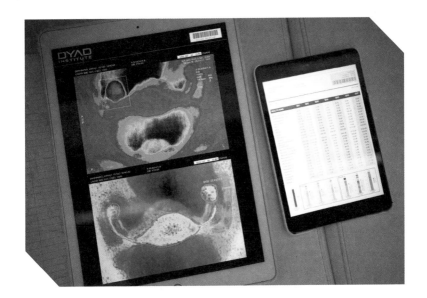

of the stomach and which contains the glands and cells that produce mucus, acid, and other important secretions — all have demonstrated the capacity to regenerate lost or damaged cells (the pancreas has shown this capacity, too!). Rachel later discusses moving forward to test Kira's ability to heal from deep tissue injuries to see if her regenerative abilities are skin deep or if, like the spiny mouse, she could heal cartilage and fat tissue. Who knows, maybe she could even surpass the spiny mouse and regenerate muscle cells.

Westmorland suggests that Kira has additional value for regeneration research by virtue of being female, because of mitochondrial DNA's role in healing and the fact that mitochondrial DNA is only passed down through maternal DNA. In the same breath, he uses this factoid to explain why, despite his (completely bunk) longevity, his mitochondrial DNA wasn't useful for Neolution's study. Mitochondrial DNA's role in aging has been the subject of lively debate in the research sphere. But generally it's been shown that mitochondrial DNA and nuclear DNA encoded factors interact and inform overall health, and that variations in genes encoded by both the mitochondrial and nuclear genomes can alter how susceptible a person is to age-related diseases. Mitochondria's connection to the energy status of the cell and to oxidative damage have tied it to aging biomarkers. If you've ever heard about free radicals and antioxidants, especially in terms of lotions, supplements, or yogurts marketed as anti-aging, the conversation is really about mitigating oxidative damage. The mitochondria might be the powerhouses of the cell, but the by-products of the power they produce cause cellular damage, which accumulates over time. Oxidative damage to proteins is mediated by reactive oxygen species (ROS), which is produced in mitochondria and which increases with age. Mutants selected for resistance to ROS have been shown to live longer than their non-resistant counterparts, but despite exhaustive studies there doesn't appear to be a simple relationship between ROS and aging. It never can be simple, can it?

Rachel also has Kira receive hormone injections to mature her ovarian follicles, since Kira hasn't yet reached puberty. The goal here is to harvest Kira's ova. What would they do with Kira's eggs? The same thing that they plan to do with Helena's babies' cord blood — isolate the Lin28a mutation and commodify it through BluZone's cosmetics lines (after first using it to grant Westmorland longevity, of course). Westmorland goes so far as to call Kira their modern-day Eve, suggesting that it would be her cells, and not necessarily the cell line that Cosima created, used in Rachel's new cloning project. That Kira would become the Original to a future generation of Neolution experiments.

WHY IT'S NEVER JUST ONE GENE

COSIMA

[...] reckless science. Focusing in on one gene when you know it's never about one gene.

(5.05 "Ease for Idle Millionaires")

As Cosima mentions when confronting PTW about his work, no biological process is ever just one gene. When scientists look for a cure to a disease, a target for a drug, or the underlying cause of an illness, 99% of the time the answer isn't as simple as identifying and studying a single gene. Westmorland's search for a single prolongevity gene is a perfect example: there have been hundreds of genes

KIRA MANNING:

A KICK IN THE KNEECAPS TO SCIENTISM

> **KIRA**
> Mommy, your sisters ... I know
> how they feel sometimes.

> **SARAH**
> How do you mean?

> **KIRA**
> Like, Cosima, when she's sad.
> Helena, when she's lonely.
> Rachel's the angriest. There's
> even some I don't know.

> **SARAH**
> I'm sorry, monkey, I didn't
> understand. Those aren't
> dreams.

> **KIRA**
> So ... what are they?

> *(4.05 "Human Raw Material")*

Let's face it: Kira Manning is a weird kid. A self-regulating gene mutation might explain Kira's accelerated healing, but it doesn't explain her other apparent abilities, namely her emotional connectedness with all of the Leda clones, even ones she's never met. As scientists, it's been really tempting for us to try and explain Kira Manning. After all, if we dig down deeply enough, if we apply the right measurements and processes, if we mine the right volumes of human knowledge, we should be able to explain her, right?

Not in this case!

Scientism, to paraphrase philosopher Tom Sorell's definition, is the view that empirical science is the authoritative be-all-and-end-all to explain life, the universe, and everything, to the point of dismissing other viewpoints and branches of learning. But Kira's seemingly psychic abilities (referred to by writers of the show as Kira's "woo-woo") are not based in science, so we're not going to try to explain them with science. And that's okay.

Any explanation that we could provide would be an unsatisfying stretch anyway. It would be like how we can explain to you that certain neurotransmitters are released in specific locations of the brain when you see something you like for the first time (for argument's sake, let's say it's a watermelon). Why did you get this response to a watermelon and not from your first time looking at a tomato? Why might the person standing next to you get no extra stimulation from looking at a watermelon but you did? These are some of the gaps in our understanding of how people think and engage, and these elements describe important facets of what makes us human. We can't explain them fully yet. Kira's woo-woo has a home in one of those gaps.

Of course, in some ways, Kira exists as a narrative device. A foil to the science that commands pretty much everything else in the show. Her abilities expand upon very human experiences like hunches and emotions and empathy. And she's a humbling reminder that, preternatural abilities aside, there are some things in our universe that humans just can't measure or explain and might never be able to.

that have ties to aging and age-related diseases and which have been proposed as candidate longevity genes. If he was trying to isolate a gene to protect against aging, he'd have his work cut out for him.

Part of the reason why it doesn't make sense to put all your chips

on one gene, as Cosima says, is that the proteins encoded by genes interact with each other, forming complexes, acting as regulators, and sometimes traveling long distances within or outside of the cells. It is not simply the action of a single protein that leads to a disease or biological process but a whole network of proteins working together.

Another reason is the redundancy within our DNA. Many proteins have multiple related forms encoded by groups of nearby genes. Ordinarily, these proteins will function in slightly different, but usually overlapping ways, either spatially (where in the body they function) or temporally (when during development they function). However, when one of the proteins is nonfunctional, either due to a genetic mutation or some environmental factor, oftentimes the other members of the group will take that protein's place, resulting in no overall loss of function. It takes the loss or mutation of multiple genes to result in a change in function and/or expression.

Additionally, a majority of the genome doesn't code for genes; many regions code for regulatory elements that control the expression of genes. The elements (known as enhancers), sometimes millions of base pairs away from the gene, when activated, associate with the genes to turn expression on or up. To further complicate things, the genome isn't simply one enhancer for one gene. A single enhancer can regulate multiple genes, and a single gene can be regulated by multiple enhancers. Nothing so straightforward as a single gene controlling a biological process exists; regulatory elements come into play, sometimes having effects on other genes, making the picture much more complicated.

In the end, Westmorland's legacy isn't exactly what he strived to achieve. He craved immortality and a lasting fingerprint on humanity, but he lived an average lifespan and died at the hands of Sarah Manning, his "flawed" cloning experiment. Clones who weren't intended to reproduce have, and clones around the world (aware and naive) are receiving cures to their genetic disease, being granted a lifespan they weren't designed to have. His work through Neolution may have taken the first steps toward the next stage in human evolution in Kira, Arthur, and Donald, and their mutations may very well be passed down to future generations of people, but that's where his influence ends. His experiments failed in the best way possible.

PROTEST

The episode titles for season five don't follow the same conventions as those for the previous seasons. They aren't pulled from iconic texts by scientists and philosophers like Charles Darwin, Sir Francis Bacon, or Donna Haraway. Neither do they tie directly to the season's specific themes, like how Eisenhower's farewell address reflects the military-science dichotomy of the Castor-Leda experiments explored in season three, or how Haraway's discussion of commodifying biology speaks to the new villains found in Brightborn in season four. The season five episode titles are taken from a poem called "Protest" by Ella Wheeler Wilcox. "Protest" was originally published in her collection *Poems of Problems* in 1914, just before World War I and at the height of the Women's Suffrage movement. (The powerful first lines of this feminist poem are often misattributed to a man — Abraham Lincoln — because #patriarchy.)

"Protest" is a rallying cry for women to find their voices and to use them. Wilcox calls for protest for a better future that isn't about the perfection that people like those in the Neolution movement seem to crave, the perfection that sees women's bodies and biology as means to their ends. She calls for protest as intervention to processes and governance that exploit these bodies, that apply violence and oppression, for the benefit of people in positions of privilege. These words speak to themes that are visited not only in season five but in the entire series. Project Leda, by its very existence, raises questions of who owns a body and its biology, especially in terms of women's bodies, and who has rights to them. Even Project Castor, used as a biological weapon, becomes a discussion of violence against women and their reproductive rights. The message in *Orphan Black* has always been clear: science and progress, even with the intent of bettering the human race, should never outstrip marginalized groups' rights to their own bodies. Humans have historically allowed for oppression in the name of progress. *Orphan Black* is a call to refuse this oppression and to push back.

THE CONVERSATION

SCIENCE CONSULTANT COSIMA HERTER AND SERIES COCREATOR GRAEME MANSON ON THE SCIENCE OF *ORPHAN BLACK*

COSIMA: We've known each other for about 20 years. We met through mutual friends —

GRAEME: — in the early '90s, Commercial Drive in Vancouver. We had eclectic friends, mostly of the non-academic, Wreck Beach variety. You were working weird jobs while going to school. *She's a barista, she works in a dog store.* Every time I would see you, you were reading advanced mathematics or Darwin. We became kind of idea buddies. I still want you to write *Dog Store*, your post-Darwin, anthropomorphic study of animal ownership viewed through the assumed intimacy of financial transactions.

COSIMA: And for much of that time, in the early days of knowing each other, we were both incredibly busy — me, trying to put myself through university by working several service industry jobs simultaneously, and you spending a lot of time between Toronto and Vancouver working in film and television. There would be a lot of time that passed between seeing each other.

GRAEME: 2007 is when I got the offer from the CFC [Canadian Film Centre] to do the Showrunner in Residence thing, the chance to develop an original concept. I, of course, thought of this vexing *Orphan Black* story, which John [Fawcett] and I had been wrestling with since 2001. I was living in Vancouver when that opportunity came up, so while I'm talking story with John in Toronto, the first person I turn to to discuss cloning science and "clones" as a concept is you.

COSIMA: I was working on my master's thesis at the time, which centered on a particular facet of Darwin's theory of evolution, but I was also knee-deep in literature regarding the political aspects of the biological sciences. Biology and the life sciences, as a formal professional discipline and practice, have always been inculcated by and thus marshaled into the service of politics, not least of which being our modern state of neoliberal drivers. Human relationships to and with concepts of life, nature, technology, sex, reproduction, and so on, are profoundly saturated with ideological and philosophical negotiations of hierarchy and control, morality and liberty, contestations between natural and artificial, determinism and chance, life and death. If biology is about life, then biology can never be anything but political! So much for this fantasy of science, and scientists, operating from some Archimedean point of pure objectivity!

GRAEME: Exactly. I knew you'd bring all these layers to a discourse — scientific, historic, social, all this context. Plus, you lived just down the street! So, some of our very first conversations were us conspiring conceptually on my porch off Main Street in Van. John and I already had a feeling that clones had been kind of given short shrift in sci-fi.

COSIMA: Clones, it seemed to me, would be a perfectly provocative way to explore and interrogate those twisted issues, especially if it could be done from a feminist perspective. I'm not sure that's what you were interested in hearing at that time! But it got the conversation started. And we've been having it ever since . . .

GRAEME: You really kickstarted the nature versus nurture conversation. You got us thinking about body autonomy, and I remember you saying very early on that "biology is always political." Basically, my super smart friend got super excited about how feminist and intellectually juicy the whole concept was, so I knew we had to be onto something. Who knew we'd still be discussing it, with an audience no less, over ten years later?

COSIMA: I remember you invited me to join you for a few days at the CFC, while you were teaching there, to talk to the class about science, biology, clones, the political and socio-economic aspects of what drives certain scientific practices and research.

Then, when I had started working on my Ph.D. in history of science, technology, and medicine at the University of Minnesota, in Minneapolis, *Orphan Black* was green-lit for production, and you asked if I would do some consultation work. Of course, I jumped at the chance. What an amazing opportunity to collaborate on ideas! Something, unfortunately, academics don't seem to do much. So, for me, the offer to write and collaboratively think through many of the ideas I'd been working on and researching for years while in academia with these amazing creatives was a chance of a lifetime.

My scholarly and academic training is in the history and philosophy of science, more specifically in the history and philosophy of biology. I am not a scientist. I have no formal training in science. But of course, I've also had to spend much of my scholarly years studying and developing some authority of knowledge about the sciences, and the practice of science more generally, that I consult on.

GRAEME: And you're an unabashed and unapologetic feminist.

COSIMA: Yes! I have always been interested in — and frustrated, even enraged — and deeply affected by how certain bodies have been subjected to and by ruling ideologies throughout history, and subjugated through various forms of rhetoric around what is "natural," thus morally imperative. Throughout history, the life sciences have serviced this rhetoric and have been used to justify some of the most atrocious and reprehensible acts of violence against great swaths of humanity and non-human life. Of course, we've also seen incredible advances in medical curatives, but make no mistake, medicine and the shifting definitions of something called "health" are hardly apolitical! The classic claim that science, and by extension medicine, services something called "truth" — as if *that's* not a contestable term! — has been adapted to fit the machinations of legally and economically articulated and sanctioned profit-making. So, it's not surprising you see, over and over again in some form or another, these kinds of issues articulated in the *Orphan Black* narratives.

GRAEME: [From a writing perspective,] I always came at science fiction through the lens of the early fiction that I read that sort of went from Albert Camus and the Existentialists on an almost direct line to Kurt Vonnegut. [laughs] I was really interested in that when I was younger, and then I was interested in world building too, again coming from novels. A novel that really affected me was *Riddley Walker* by Russell Hoban, which is set so far into an apocalyptic future that language has devolved, so [the book] has its own language. And then even earlier, there was the kind of apocalyptic sci-fi movies from when I was a kid. I always had an existential view of the future; I always felt, since I was a kid growing up in the Cold War, that I would see the breakdown of society as I knew it. So, I've always been interested in what that would look like, in dystopian futures, and so I just remained interested in sci-fi, and it just kept being more and more imaginative and interesting. And I was super affected by movies like *Brazil* that added dark humor — the black humorous aspects of existential crises, that's what I like. At the core of sci-fi that gets to those big unanswerable questions like *who am I, what are we doing here, when do you start or cease to exist being a human.* I found those to be the most interesting questions because there's no answer. So, you're constantly writing to questions, writing to mystery, and I always find that if you can write your way to the end of a mystery, and that mystery ends with a larger question, that's a story well told.

COSIMA: I agree. I prefer the dystopic kind of storytelling as well. It seems that biological sci-fi is what you write, as opposed to physics sci-fi — there's also sci-fi dystopia with time travel, monsters . . . But I wonder did you just fall into the bio-horror sci-fi or if that's a strain of sci-fi that appeals to you?

GRAEME: It really does. That strain of sci-fi does appeal to me. And John Fawcett was very influenced by it as well. And I'm talking about the sort of David Cronenberg school of bio horror.

COSIMA: Yeah!

GRAEME: We both loved that stuff. John, on a visceral technical filmmaking storytelling side of it, and for myself for the big questions of it. Those big existential questions again that Cronenberg always approached so well and in this slow, creepy, bio-horror kind of way. You know I'm not that interested in space epics, or physics, in that kind of thing. Though I love to watch time-travel movies, good ones, I hate to write them.

The premise of *Orphan Black* demanded a deep, deep understanding of biology and where biology was — *is* — at this point in time. And in my earliest conversations with you, I began to see that biology was the new physics, that it was *the* science of the 21st century. That all of the big questions about what makes us human were going to be encapsulated in biology within the next 50 years. We're going to face all of that.

COSIMA: We are facing that.

GRAEME: It was an inexhaustible source of concepts. And science on the verge, science on the cusp of revelation or discovery — that proved to be super ripe for *Orphan Black*. As a world to explore but also thematically in a show about identity and ownership, biology became the thing. It became so clear that these were the deep themes that would make the show sort of perpetual — that we could mine those themes of nature versus nurture in a way that, to me, just suggests drama and conflict and story. That it's a great engine.

COSIMA: I know that you originally envisioned the main character — who turns out to be Sarah — the hero, as a woman, and I know that many things have budded and bloomed since then. Do you think there's any particular benefit or strength in telling it from a female's perspective? I am curious about the politics, and what your journey through understanding that might be, but also in terms of the biology, the science of evolution, and, you know, bodies, how you see having a female hero aids in the storytelling and the drama. I mean, it could have been told from the male perspective with a bunch of male clones. That wouldn't have been unusual, but you actually chose to take a very, very different turn on it altogether!

GRAEME: It was a conscious choice from the beginning. When we first were conceiving of *Orphan Black* from that opening scene of "Okay, so, a woman gets off a train and sees someone across the tracks and in that moment their eyes meet and the person across the tracks, who's a doppelganger, commits suicide" — from that we were asking ourselves, *Who was that person?* It was always a woman. Even in that opening pitch, it was always a woman. But it's interesting that at that time we knew we wanted an ass-kicking underdog hero and that was the extent of it. At that time, it was like, "Oh yeah, of course, we want an underdog; it better be female driven." As an underdog, right? John has often liked women at the center of his stories. But as I looked into cloning, and as I worked with you on what it means to clone, it was *logical* that you would create females. And that immediately made it a feminist drama, because it immediately put questions of body ownership at the fore. Who owns your body is one of the central questions of the show, and a central issue of feminism. Exploring it deeply has greatly affected how I approach story, female stories; it's affected how I see the business politically and has driven me to want more female writers on the show, and more female writers in the business. It has affected me as a feminist male.

COSIMA: I have definitely observed that. Do you think that there is something also very interesting and profoundly informative when you look at biological sciences as they are marshaled into the service of politics? How that affects body agency, who we get to be in the world . . . Has working on this particular kind of science-fiction helped you recognize how political that science actually is?

GRAEME: Yeah. The commodification of science and the explosion of junk science, the obfuscation in science. Science and biology is always political. That's something that you've always said, that I've learned. Without a doubt, it's opened my eyes and made me, I think, pretty pessimistic about how science will be marshaled in the next hundred years. Ownership, patents, conglomerations — the opposite of diversity is what is happening in the commodification and the corporatization of science. Particularly in pharmaceuticals and particularly in agriculture. It is removing diversity, and it is removing our choice, and our freedoms. Technology and science are being used against us as political tools, and they will be used in levels of entitlement, levels of who affords it. It stratifies society even further. Top people get top things. Top people will get the life-extension science that is coming down the pipe. Poor people won't.

COSIMA: Let's talk about some of the particular kinds of science that we explore. Is there any particular science that we explore, for any reason, that's your favorite? Like, you *really* like looking at patenting, or you really enjoyed thinking of evolution in terms of chance and contingency, or clones or . . . What do you really dig?

GRAEME: What I really dig is the spark of Sarah. The spark of Sarah is that amid all of this marshaling of the science, amid all of this control, all of this projection of power, to create a world in your vision or that serves you, that Sarah is chance, Sarah is contingency. That nature in its diversity can adapt, can mutate, and it can bring you down. And that is what will happen to mankind.

COSIMA: Yeah. It can also compel you, propel you forward. That mutation is the spark that allows us to adapt, right?

GRAEME: And, you know, that's interesting the way that that also works itself in an extremely complex world in chaos theory, how chance and contingency bring down systems.

COSIMA: Because you can only design so much. There are only so many features you can understand, let alone account for and design.

GRAEME: Right.

COSIMA: It is an interesting commentary on design, as opposed to indeterminacy.

So, I approach my academic study of some of these issues very differently than we do on the show. I've had a really sharp learning curve because there are certain kinds of accuracies that have to be sacrificed for telling a story on television. Maybe you could talk a little bit about that process — how particular kinds of scientific accuracies have to be molded, made a little bit plastic in order to tell a narrative. How plastic do you want to make things to service a story? Some things you'll make plastic to serve a story, and some science points benefit from *not* being made so plastic.

GRAEME: It's a balancing act. We've always been interested in *okay it's a grounded sci-fi*. So, we want to deal with the science in the here and now. But it's the science that you can imagine in the world that we know now, and the science that we know now, all of the facts, that's behind the lab door in the large corporation — what's going on there? What's one step ahead? What is one step ahead of what the public knows the science is? That's the interesting jump-off point. We conceive within the realm of existing science, and then we ask ourselves, "What's the jump-off point?" Things like the timeline get compressed. How long science takes. But I really dislike science shows where every act is a scientific discovery, every plot point is a eureka moment. Because science doesn't work that way. Your eureka moment should only spawn more questions. That's a certain reality in writing sci-fi. You don't find the magic element to put in the warp drive that gets you out of the vortex.

COSIMA: [laughing] You don't find the magical mutated gene, either! Or if you do, you don't know what to do with it!

GRAEME: Right, you don't know what to do with it! Or you use it and it creates more problems. Which is one of the foundations of the clones is that they pushed ahead in this science and they messed them up.

COSIMA: So many inventions are actually made by accident. So many incredible things we understand about the world through science actually happened by accident.

GRAEME: It's hard when you're working with plot, when you're trying to push these stories forward, because writers grab on to revelations to hang story on. When you look at *Orphan Black*, over the course of a bunch of seasons, there are actually very few scientific revelations. There's three or four a season. And usually it's, you know, *oh my god we're patented*. Well, that creates a whole raft of questions. Or discovering germline editing, and looking at that. That's a contemporary science that we thought was rife with future possibility for good and evil. But often those things that you're trying to hang your hat on when you're telling a story, those scientific concepts are better to be treated in a general fashion than to treat them like a eureka moment, like they're an answer. If you treat them in a general fashion, you treat them as: *what do they mean? what do they mean politically? what do they mean to that character? what do they mean as a concept?* And when it comes to biology, again, you get back to *who owns it? Who owns you?* That existential question.

COSIMA: Is there anything that we've done that you wished we could have explored more but the platform of television narratives didn't really allow us to explore more or go deeper? So, keeping in mind that we're writing TV, we're not trying to answer deep scientific or philosophical questions, because we don't have the capability for that. That's why there are thousands of researchers trying to understand particular kinds of evolutionary mechanisms. We don't position ourselves to answer those things, but there are some things that we look at that we kind of didn't get a chance to dig deeper into. For me, like, the patenting — that's great, but there's not a lot of real estate in the narrative to go into it so deep.

GRAEME: My favorite scene in all of *Orphan Black* still to date is the end of season two, I believe, when Sarah and Cosima lie in bed and hold hands and they talk about Buckminster Fuller and Cosima's nautilus tattoo and they talk about the Fibonacci or the golden mean, I guess, right?

And, I mean, to me that is a triumphant scene because it's distilling a whole bunch of science down into a feeling and a tone of mystery and understanding and beauty, and it's really hard to get characters there — to fight them through all this science and plot. I wish we could have had less plot and more depth. Because we move so fast. I would have liked to slow down and do more scenes like that, that examine the bigger questions, the deeper questions. The problem with pursuing a mystery is that people want answers. I love mysteries without answers.

COSIMA: Yeah, me too.

GRAEME: Take Kira's *specialness*, quote, unquote, for instance. It's something that I have had to explain and re-explain again and again and again and again to smart people, because I don't want there to be an answer. And people wrack their brains. And I'm like, *What makes Kira special?* This is not her biology, okay. What gives her that connection with the sisters, that empathetic connection? It's not explainable by science. It's the most human thing about her. To me, that's the most simple and beautiful and elegant thing to want to put on screen, and conceptually people can't wrap their heads around it.

COSIMA: Because it's not explainable by science.

GRAEME: No! It's a beautiful little moment — I want to get it to a beautiful little moment where it all comes down to you know what? All of this running around, all this science — the most important thing about this whole show: you can't explain that little girl's special powers.

COSIMA: It also speaks to a human being as more than the sum of their parts. There's something in there as well. If particular kinds of scientific methodology demand that questions be set and answered in a sophisticated, sequential kind of way, then, ultimately, you're only going to see the organism as the sum of its parts and nothing more, right? I like that we actually explore that as well. I've told you this before, but one of the most useful and profound pieces of advice for me was from my old mentor and dear friend John Beatty. When

CLONE CLUB Q&A

What's the science behind Cosima's nautilus tattoo?

COSIMA

So, this spiral, this is the golden ratio and it's a mathematical pattern that just repeats itself in nature, in flower petals and honeybees, and you know, the stars in the galaxy, and in every molecule of our DNA.

(2.10 "By Means Which Have Never Yet Been Tried")

The golden ratio is a mathematical concept discussed by Cosima and Sarah in the season two finale, "By Means Which Have Never Yet Been Tried." Cosima mentions it in reference to the nautilus shell tattoo on her wrist, but she keeps it brief for Sarah's sake. However, it is quite an interesting subject that warrants a tad more attention.

Two numbers are in the golden ratio if their ratio is the same as the ratio of their sum to the larger number. In other words, if $A > B > 0$, and $A/B = (A + B)/A$, then the numbers are in the golden ratio. The Fibonacci sequence is a sequence of numbers in which each number is equal to the sum of the two previous numbers in the sequence. The first ten numbers in the sequence are: 1, 1, 2, 3, 5, 8, 13, 21, 34, 55. If any number in the Fibonacci sequence is divided by its immediate predecessor in the sequence, the quotient is equal to the golden ratio. A golden spiral is simply a spiral that has a growth factor of the golden ratio — the spiral gets bigger by the value of the golden ratio every quarter turn.

In nature, the golden ratio is found in the branching patterns of trees, in the veins of leaves, in honeycombs, in spirals — it's pervasive throughout so much of the world around us. As for the nautilus shell, such as the one on Cosima's arm, the spiral of the shell is actually a modified version of the golden spiral, growing by the golden ratio every 180 degrees instead of every 90 degrees. But it is still connected to this equation and sequence.

Many artists model their work after the golden ratio because shapes that conform to the golden ratio are considered to be more aesthetically pleasing than those that don't. Maybe that's why spirals in nature seem so beautiful! The golden ratio is also seen in music, architecture, design, and basically every aesthetic you can imagine. It's amazing how much a simple math equation has come to permeate our everyday lives.

Fun fact: November 23, or 11/23, is known as Fibonacci Day and is, coincidentally, *Orphan Black* cocreator Graeme Manson's birthday. Cosima would approve.

I was working on my master's thesis, he said, "You know, Cosima, there are two ways to approach a discourse or a domain of knowledge or enter into a conversation: one is to find a question and set out to answer it. And the other way is to actually look at it and set out to problematize it." So, when you talk about speaking to questions more than answers, that's an approach I really appreciate. And I think that's something we have always done with the science of *Orphan Black*: problematize it. Problematize the assumptions people have about evolution, about women, about your psyche or about psychology, or about our fatalist determinism in the world. We open those questions and pull them apart.

GRAEME: I'm not even sure that I know, intellectually, what to "problematize" actually is, but I think that that's what I do as a writer every day.

COSIMA: Yeah, you do. You absolutely do. [laughs]

GRAEME: A story is a great big freaking problem that you have to figure out how to tell. And telling it in TV or film is telling something absolutely completely different. It has to sift through so many hands to come out the other end. It's a very distinct and problematic form of storytelling.

COSIMA: It's like having a grad committee [laughs] — everybody's got to comment on your thesis! Whether it's actually relevant to your thesis or not.

Coming from a completely different background of communicating scientific ideas or philosophical ideas, I've really had to learn what can work for television. To learn how to "consult" has been enormously profound and laborious, because it's unlike anything I've ever done. An academic would not describe certain concepts in these ways and, in fact, would be offended that so much gets sacrificed or reduced into a single point.

GRAEME: Part of that battle is to reduce them to an understandable point, yet communicate their concepts so that they have bigger meaning, broader meaning.

COSIMA: There's a whole line of people that we have to communicate particular kinds of ideas — scientific, philosophical, and *political* — and then justify why you want to incorporate those into a story. And some of the ideas are quite contentious. We're talking about patenting; we're talking about reproductive rights; we're primarily talking about the agency of oppressed human beings and bodies in the world, not just women, but that's one of the vehicles. Biotech and body-mods . . . You really have to defend certain things that some people might find offensive — and then incorporate them into a story! Is there anything in particular that was tricky to communicate to all the various people involved in producing a TV show?

GRAEME: I think it all just becomes part of the same story. I feel like I have to defend literary character choices, any other choice, just as much as the science. The science has its own challenges in making people understand it who are non-scientists. But I'm a non-scientist, so I have to understand it myself before I can communicate. Yeah, there are challenges. But we kind of want the same things out of the science. Though John isn't as deeply interested in the science, he's very good at making sure it makes sense.

COSIMA: He is really good.

GRAEME: I don't like lines and lines of techno jargon, but there's a test with John. If his eyes cross after a scientific exchange of more than four lines, it's too long or it's not clear enough. And that helps us get clear science.

COSIMA: It's a good litmus test.

GRAEME: And I want it to be clear, too. There is tension there, though: I like to communicate bigger concepts.

COSIMA: Each season we keep the main action-conspiracy going — that's not something I necessarily have any hand in — but each season we move from one sort of large domain of science, and social-political-philosophy questions that we buttress with the science. Not unlike many sci-fi writers, we use science as a vehicle for exploring and interrogating different kinds of social and political issues. The integrity of the science is there throughout every season, and we use it as the spine around which we spin the action-story, the "run and jump" aspects of the characters' lives and activities. We use the science as a way of embodying various controversies and rubrics of biotechnology and bioengineering, genetic determinism and genetic manipulations, the industrialization and monetization of the biological sciences, family relationships and heredity, personal agency, freedom, autonomy, the ways that the discipline and practice of biology aids the oppression and disenfranchisement of certain populations throughout history, and health, medicine, reproductive rights, the notion of the individual, and most broadly what is life.

For example, for the first season, we had many conversations about some of the basic principles of (Darwinian) evolutionary theory; contesting visions of contingency, chance, and determinism; individuality; identity; (autonomous) selfhood; what it means to have (or not have) ownership of one's own body; and bodies, tissues, cells, genetic lineages as forms of property and profit. We looked at these in many, often mind-bending, ways, so you see that explored and exploited in that season.

Season two, we explored the long, complicated, historical relationship between science and religion, and the social practice of science. Eugenics. Control over one's own body, and the existential choices that direct one's own life. Sexual abuse and the use of women's bodies for misogynistic socio-economic gain, especially in regard to reproductive rights and the commodification of women's bodies. Sisterhood — both in the genetically related sense, and the chosen sense.

Season three was guided by questions of authority, homogenization of a population according to the violent rubrics of white-Euro superiority, masculinity and aggression, how those ideologies have colonized concepts of self and body and expectations of identity; the insertion of different arms of industrial funding of scientific research (like, of course, the military) and how that affects the general praxis, thus the outcomes, of scientific research; contemporary aspirations of patriotism, nationalism, and mass eugenically genocides; bioterrorism and bio-power; siblinghood and family.

Like, for example, as you pointed out, in season three, we encounter the Castor clones, the literal, genetic brothers of the Leda sestras — but this relationship of blood (i.e., "natural") siblinghood is confounded by the fact that it has no emotional, or so-called spiritual, meaning between the Ledas and the Castors. The notions of chosen family relations were really important to you for this season.

Season four was predominantly situated around the difference between science and technology/engineering. Where, for example, science aspires to speak to and uncover something called truth, and engineering aspires to a construct that is simply a sum of its parts, we look at a body as a kind of machine that can be "fixed," improved on, rebuilt according to particular specifications for order; the commercialization and profit-driven machinations of industrialized reproductive capacities; the aesthetics of designer babies; the limitations (and the public's hopes and expectations) of gene therapy, and germline manipulations; socially constructed values of desire and how that affects consumer-driven choices; the tangled mess between definitions of scientific success and failure; the unreliability of technology and technological interventions into living systems that are barely understood; and, of course, issues of consent.

Season five is a long-view discussion of not only what we wove throughout the narratives in the past seasons, but also a launching point into the future that is unwritten — not simply unwritten by the *OB* writers, but unwritten in the sense that it is impossible to know, let alone predict, what the future may hold. The future is an undetermined, unpredictable, and contingent temporal-spatial environment that we really have no control over — despite our best efforts and most concerted attempts to control it. Adaptation and resourcefulness are guiding themes here.

But one of the most difficult and important things we tried to integrate here, using longevity and immortality science, was to throw patriarchy and the proliferation of long-ruling ideologies into stark relief. Prolongevity science and the aspirations to live forever, its hubris and its faults, formed the architecture by which we explored the institutional, systemic, and individually embodied regimes of historical, patriarchal ideological power that has oppressed, subjugated, violated, controlled, and prevented diversity, equality, and solidarity. We used longevity science as a stand-in for the way ruling regimes mutate and morph throughout history in pernicious and insidious ways that fracture, atomize, and disenfranchise populations of humanity. As an allegory for the need to resist, protest, fight against deeply established, long-standing ideologies that make us feel powerless to control our own lives, and hope for a better future.

In terms of how these big ideas are woven into story, maybe you can talk about how our process and the process in the room with the writers.

GRAEME: First, I had initial conversations with John about big-picture stuff and gleaned some of the things that he was interested in. We generally have a thrust and a concept, and we're looking for a new angle of science, or a new branch, to attach to that story. Not necessarily to be the story or be the plot, but it becomes plot because some of it inevitably becomes goal-oriented in terms of discovery. In the beginning of the year, after those conversations with John, that's when you and I really tuck into some bigger-picture stuff. Pretty much every year we've done that. And then I take that, plus the time I've spent with John, and that's the first stuff that I take to the writers' room. It's a package of work with John about story, and then it comes with a package of ideas as well that are connected to the science. The first job of the writers' room is to look at the big picture every year. Always having to look one season ahead to where we're going. So those early days and early season writers' room, a good three to four weeks, are a conversation about big story ideas as well as the concepts — the scientific concepts we're trying to communicate and how they might play into story and buttress character, and how they reflect overarching series themes of identity. We always hold it up to that. In doing so, usually, we end up picking that branch of science — be it epigenetics or germline editing, or —

COSIMA: — where the funding comes from?

GRAEME: Yeah, or patenting. Gene data collection.

COSIMA: Yeah, I love that stuff! [laughing] I cast a really wide net, and I start researching all kinds of things I think might be interesting, and I bring to you a number of different ideas that I try to convince you are really excellent to look at, and you filter some of those out. You keep the ones that really appeal to you, the ones that you know are really usable to motivate and to mobilize story.

It's been an evolving process. I knew relatively nothing about television or how television is made before *Orphan Black*, so our process has also included a steep learning curve for me. If I were to sum up, in broad brushstrokes, what I "do," it might be easiest to describe it as basically curating ideas, and inspiring you to think about ideas you've never thought of before, in ways you haven't before. I research, prepare, and present political, scientific, historical, social, and philosophical material for the writers to think about, and offer ways to think through these ideas relative to what could become an exciting and provocative character-driven narrative.

I think one of the most remarkable things about whatever process we might have is that, in our conversations, you will ask a few sharply pointed and provocative questions, and I will often run with them . . . Like, I mean run off at the mouth for indeterminate amounts of time about whatever idea it is we're interested in trying to work through. And, for some reason, you seem to get a lot out of that. I haven't a clue how you endure it! You're a great synthesizer of information — you can take in an enormous amount of material I give you (written, verbal, visual), distill, and synthesize it into a coherent, embodied narrative. That is an extraordinarily difficult task!

One of the aspects of our process that I think makes us really jibe together is the fact that, as thinkers, we both have a penchant for thinking and writing toward the questions — questions that provoke and inspire, confuse and confound, vex and perplex, questions that may or may not have answers. We write to the mysteries, the ineffable and often invisible assumptions that underlie and incite human activity, without trying to answer those questions. This is something in particular that I love about working with you.

And, last but not least, our process includes a *lot* of falling-out-of-our-chairs laughter! Laughter. I'm a very earnest person and not particularly on-the-spot witty. You're one of the funniest, wittiest people I know. You laugh at me a lot, and you make me laugh a lot. Your humor is inspiring.

One thing I've really wondered — and I already know your favorite scene — but I'm curious to know, is there any particular science, any idea that you think is really, really cool and you want to explore, but is super hard to integrate, and one that we integrated that you think is really excellent? Like epigenetics, as an example. That's a really difficult concept to wrap your head around.

GRAEME: The interesting thing about epigenetics is in the years that we've been making the show, epigenetics is now a thing. It's burgeoning, but we chose not to pursue epigenetics in the first year because it was too woo-woo.

COSIMA: It was too complicated to communicate. But is there anything in particular, from reproductive technology to whatever, that you had to really struggle to understand, that appealed to you on both a visceral and an intellectual level? There's lots of concepts that are new.

GRAEME: All of the work that we put into how Kendall Malone could be Mrs. S's mother — and kudos to Chris Roberts, who should be mentioned as someone who took a strong lead role in the science of *Orphan Black*. I can't remember how we got to chimera, but it was how the whole thing could be possible in a moment of dramatic revelation. That was a doozy.

COSIMA: I guess that's the term — having the idea and then turning it into something that could be a "dramatic revelation." Especially a few years ago, I didn't really understand how the information needs to come to you in such a way that you can actually create a dramatic revelation with it. That's something that I had to learn. I would be staunchly not wanting to move on a particular idea, because I didn't want it to be perverted by it being so reduced. So, I've really had to get over it, and become really humble; there are some things I won't move on, because I just know it's to your benefit to not make it too fantasy. But there are other things, where, again, it's making it plastic so you can develop a narrative about it. That's been an interesting working and learning process.

GRAEME: You can do that by making it more than the formula.

COSIMA: And that's not even what I enjoy about looking at science anyway! Yeah, the formula, equation — I figured out the letter x! Sometimes we have to figure out the letter x in the equation, but other times we don't.

One of the things I thought you integrated in a way that was quite beautiful, and I still feel really proud of, especially when you're talking about overarching themes of identity and individuality, and questions of agency, especially in season four, is the conversation between Susan and Cosima about consent. In that scene, there are so many levels that we're broaching. But in terms of the science, in terms of how we explore questions, how we investigate, what gives us the rights to investigate . . .

GRAEME: The broad examining, the broad issues of ethics with Cosima has been a real journey. I think I know a lot more about ethics, I think I'm more ethical now, after five years with that character.

COSIMA: Good, I'm glad! [laughing] I'll take that both as a comfort and a compliment! I think that we have carried some of those questions — what is science, how is science done, why is it done that way, what does it give us if we pursue certain kinds of questions, or investigate or research particular strains of thought? Why do we need to uncover the secrets of your biological being? Those are the things that really appeal to me, and using historical examples that we model it off of — and suggest this isn't just fantasy — these are actually things that go on in the real world in some way or another. I've had such enormous admiration and respect for how you've been able to carry some of those deeper questions and weave them through conversations amidst all the "run and jump and monster" sort of activity. I think that's been really successful.

So, this is the end of the show — is there anything in particular that we never got to that you wish we might have been able to? Because we've gone through lots of ideas, many that get discarded — they're either too complicated, too offensive, too unknown, too unworkable, or we just don't have room, even if we loved them. But were there any ideas that you're like, "You know, we never actually wrote a scene . . ."

GRAEME: There is one thing. I wish that we could've expanded our ideas about identity more — and then there's a technical thing: I wish we had more time to shoot more clone scenes. The limitation of how many we could do per episode limited how deep we could go with them —

COSIMA: — and how we could show it.

GRAEME: Yeah. Sometimes it can be a hard battle to do a clone scene where nothing's happening — as in no great technical wizardry, where there are two clones having a conversation — like the Cosima-Sarah scene in season two. Those conversations or those concepts play better in clone scenes. It's the Cosima character talking to our supporting cast, which are fantastic — they're the lifeblood of the show. But having those conversations face-to-face — we were extremely limited in how many of those we could do. That sucks.

COSIMA: Yeah. So, you know I have very strong feelings about identity and identity politics, diversity in the world, and equality based on diversity. I look at it politically and socially; I'm an unabashed feminist. I also look at it through the history of evolutionary theory; we actually cannot homogenize species to the point where we can't cope or can't adapt. Diversity and mutations, that which is not like the other, are absolutely necessary. Biological sciences are a very good way to compel that conversation, without necessarily using it to answer. But what is it to you, personally, that makes identity so important?

GRAEME: I think it's more about writing; I think it's more about writing characters that I like. You just have to be constantly interested in people's identities. To understand and write a deep character, you have to know how they see themselves, and how they trample or glide through the world. I think we're constantly dealing with questions of identity as writers, and it's a really rich theme, not only literarily, but it's cinematically super-rich, which is the goal that John and I landed on together in this concept of cloning and identity. It was a really potent mix.

COSIMA: One final thing people often want to know about: how much of me is in the character of Cosima. And I don't know! I can't really answer this — it's too difficult for me to step back far enough to actually see what nuance may or may not be there insofar as how close Cosima N is to Cosima H. I didn't have any hand in writing or directing that character. I can say that Cosima N is a fictional character, so whatever similarities there may be, they are most certainly mitigated by fictionalization. Like, for example, I'm not a scientist. I don't have, nor have I ever had, dreadlocks. My parents did meet and marry in San Francisco and were American drug-culture and anti-war protestors straight out of a bad Hunter S. Thompson essay, but I was born and raised in Canada. I've never lived in Berkeley. I did spend most of my adult life living on the West Coast of Canada, however, so I've no doubt some of that lifestyle has infiltrated my own personality.

GRAEME: Because you were consulting and involved in the conversation so early on, we just started calling "the tech clone" after you. You know, the geek member of the squad, who stays in the lab and solves problems via *science*. But, of course, we were trying to make the character singular. I always admired how Real Cosima's empirical approach to truth mixed wonder and humility, and left a healthy space for mystery. And I really liked this character idea of a humanist, super earthy, scientifically brilliant hippy. That's *really* not Real Cosima at all, but Buckminster Fuller sure is, and you chewed my ear off with Bucky. So, I don't really equate the Real Cosima with the Cosima I see onscreen, but she feels like people we know or admire, so she's both of us.

There was a moment, right after we got the green light for season one, I told you straight up, if you wanted to change the character's name, this was the time. Your response was hilarious; you said, "Don't you dare steal my post-modern moment."

And it was a narrative from then on, you know? We cast Tat. Who knew? I never said anything to Tat about Real Cosima versus her character Cosima. I just said, "You have to meet my friend, who's going to stir up your brain." So Cosima met with Tat early on to talk science. In fact, you held an open "Cloning 101" seminar for any cast or crew who wanted to attend. Outside of the story department, you did these kinds of conceptual meetings often, particularly with the art department. You and Tat had your own thing going, too. With these feminist flags planted, and all these characters with complex delineations of perspective, you helped Tat sort out a lot of big questions, under a lot of pressure. Your thing was really about honoring every character's point of view.

But it must be said: Real Cosima is a hand talker. Like amazing. Your hands sort and categorize and finesse us to understanding. I think we can safely say Tat took on a bit of that animation in her performance. But that's really less about physicality and more about Tatiana capturing Cosima's enthusiasm for concepts, theories, and the big ideas that give her goosebumps.

GLOSSARY

ALLELE: One of many alternate versions of a gene or gene locus (the location or position of a DNA sequence on a gene) when more than one version exists.

APHASIA: Difficulty understanding or producing language as the result of a lesion affecting at least one of the speech-related areas of the brain.

BASE PAIRS: The building blocks of DNA: a base pair is a unit of two nucleotides (adenine and thymine; guanine and cytosine) held together by hydrogen bonds. Nucleotides are the letters that form the code of a DNA sequence. If you imagine the DNA double helix as a twisted ladder, the base pairs form the rungs of this ladder.

BEHAVIORISM: The theory and belief that human behavior is a result of environment and learning rather than genetics.

BLASTOCYST: A blastocyst is a ball of roughly 200 cells that develops about five days after fertilization and later becomes an embryo. It possesses a hollow cavity and an inner cell mass (ICM), the latter of which can be a source for embryonic stem cells.

CHIMERISM: A genetic condition in which one organism contains more than one cell line. In humans, this can occur when one twin absorbs the other during early fetal development in the womb.

CLUSTERED REGULARLY INTERSPACED SHORT PALINDROMIC REPEATS (CRISPR): CRISPR are sequences of DNA in the bacterial genome that are part of its defense system, allowing the bacteria to destroy sequences from viruses. Paired with the protein Cas9, CRISPR can be used as a genome editing technique by disrupting or modifying sequences on target genes in multiple species.

CUMULUS CELLS: A cluster of cells that surround the oocyte (developing egg cell) while it's in the ovarian follicle, as well as after ovulation. Cumulina the mouse, the first mouse to be cloned from adult cells, was created using the nucleus from a cumulus cell.

DETERMINISM: The theory and belief that human behavior is inherited, a result of genetics.

DISCORDANCE: In genetic studies, particularly in twin and sibling studies, discordance describes when a trait appears in one individual but is absent in the other.

DIZYGOTIC: Resulting from two separate zygotes. Dizygotic twins are often referred to as "fraternal twins."

EMBRYONIC STEM CELLS: Undifferentiated cells retrieved from the inner cell masses of blastocysts or from embryos. These cells have the potential to become almost any kind of cell.

ENUCLEATION: The removal of the nucleus (the part of the cell that contains genetic material) from the cell. This is an important step for somatic cell nuclear transfer.

EPIGENETICS: External or environmental factors that are not part of the DNA sequence that affect how a gene is expressed.

GENE THERAPY: A method for treating disease by delivering genetic material directly into the patient's cells, changing the expression of a target gene.

GENE VECTORS: For gene therapy, vectors package and deliver therapeutic DNA into the patient's cells. Often gene vectors are viruses.

GENOTYPE: The genetic code that indicates what phenotype, or visible traits, an organism will express.

HEMIPARESIS: Weakness on one side of the body.

HYPERMETHYLATION: An increase in methylation at cytosine and adenosine residues (two of the four basic building blocks of DNA). These additions reduce gene expression.

LESION: In the context of brains, lesions are an area of tissue damage in the brain. A lesion can have many causes, such as traumatic brain injury, cardiovascular events (such as stroke or hemorrhage), infection, or tumor.

MEIOSIS: The process by which reproductive cells divide to produce gametes (sex cells, such as egg cells or sperm). Human cells produced by meiosis are haploid (containing one copy of chromosomes) and have 23 chromosomes.

METHYLATION: The addition of methyl groups to cytosine and adenosine residues on DNA that function as epigenetic markers.

MICROBIOME: The collection of microbes (bacteria, fungi, protozoa) that live in and on the human body and influence health and body functions.

MITOSIS: The process by which somatic (non-reproductive) cells divide to produce daughter cells. Human cells produced by mitosis are diploid (containing two copies of chromosomes) and have 46 chromosomes.

MONOZYGOTIC: Resulting from a single zygote. Monozygotic twins are produced from one zygote that splits and are often referred to as "identical twins."

MOSAICISM: In genetics, the presence of more than one cell population with differing genotypes in an organism where all populations are derived from the same fertilized egg. Contrast with chimerism, where different cell populations in a single organism arise from separate fertilized eggs that have fused.

NEOLUTION: Neolution is both a fictional philosophy and a fictional organization with a focus on using scientific research and novel technologies to transcend human limitations and work toward self-directed evolution. Many of its projects are ethically dubious at best and definitely eugenic. Neolution is the main antagonist in *Orphan Black*.

PHENOTYPE: The visible traits that an organism has, coded by their genotype.

PLURIPOTENT: Describes a stem cell that is able to give rise to a majority of cell types. (The only cell type a pluripotent stem cell cannot give rise to is a trophoblast cell, a cell type in the blastula that becomes the placenta after implantation.)

POLYMERASE CHAIN REACTION (PCR): A technique for multiplying a small amount of a particular DNA sequence to produce a sample of thousands or millions of copies of that sequence. This technique is useful for DNA sequencing and disease diagnosis.

PRION: In its infectious form, a protein that causes disease by triggering other proteins to misfold, form clumps, and cause damage to brain and nerve tissue. The most commonly known prion diseases are spongiform encephalopathies like Creutzfeldt–Jakob disease, kuru, and mad cow disease.

PROGENITOR CELLS: Progenitor cells are similar to stem cells in that they can become one or more different types of cell, but they are more limited in terms of what types of target cells they can differentiate into because they are already lineage-specific (often it is only one type) and cannot self-renew indefinitely like stem cells can.

REPRODUCTIVE CLONING: Cloning for the purposes of creating a living cloned organism. The clone embryo is implanted into a host womb and develops to term. This form of cloning is illegal in many countries.

SCIENTISM: The view that empirical science and the scientific method can explain everything, to the point of dismissing other viewpoints and branches of learning.

SHORT TANDEM REPEAT (STR): A short sequence that repeats hundreds of times in a row on a DNA strand. STR analysis is one of the most useful ways of comparing specific sequences on genes from two separate samples.

SITUS INVERSUS: A condition in which the internal organs are located on the opposite side of the body than is normally observed. Dextrocardia is a specific form of situs inversus that Helena demonstrates: her heart is located toward the right side of her chest, whereas the other clones (Sarah included) have their hearts located on the left.

SOMATIC CELL NUCLEAR TRANSFER (SCNT): The technique by which clones are made! The DNA-containing nucleus from the organism to be cloned is removed and placed into an enucleated donor egg cell, which is then placed in a nutrient broth, given a little shock of electricity, and left to divide as a fertilized egg normally would to form an embryo.

SPINDLE APPARATUS: The spindle apparatus is primarily a mechanism of microtubules — built from proteins — that form to separate genetic material during cell division so that the two resulting cells have the same number of chromosomes. This apparatus can pose challenges to SCNT and was an issue for both the first generation and the Charlotte generation of Project Leda.

TELOMERES: Regions of repeating sequences at the ends of chromosomes. The shortening of telomeres has been associated with aging and age-related diseases.

THERAPEUTIC CLONING: Cloning for the purpose of developing treatments or therapies derived from the cloned embryos. This form of cloning does not result in a living cloned organism.

TRANSHUMANISM: A philosophical movement that seeks to transcend humanity and reach the next level of human beings, often through biological experimentation and body modification. The inspiration for the Neolution movement in *Orphan Black*.

X-INACTIVATION: People with the XX genotype, like the Leda clones, have two X chromosomes, one from each parent (in the clones' case, both X chromosomes come from Kendall, whose own X chromosomes came from both of her parents). To prevent having twice as many X chromosome gene products as needed, one of the X chromosomes is randomly inactivated. Which X is inactivated is not necessarily the same across all cells of the body.

ZYGOTE: A fertilized egg cell.

ORPHAN BLACK AND HISTORY

A TIMELINE

The basic conceit of *Orphan Black* is that humans were successfully cloned in the 1980s. While this didn't actually happen (as far as we know!), the techniques used to create the clones in the show are borrowed from real life. Much of what we see on *Orphan Black* is based on real-world historical events, experiments, and scientific discoveries. The scientists on the show just seem to have figured things out a little bit sooner than real scientists — and to have applied their discoveries differently. See where *Orphan Black*'s alternate universe fits into and interacts with our own in this timeline, which includes events that occur or are alluded to in the show as well as some important real-world events that give context to our understanding of the science that we see on our screens (or read in the comics).

LEGEND: REAL LIFE • **ORPHAN BLACK**

1600s

| 1620 | Sir Francis Bacon's *Novum Organum Scientiarum* is published |

1800s

1818	*Frankenstein* by Mary Shelley is published
1842	**Percival T. Westmorland is born**
1856	Mendel starts his experimental garden to study heredity in pea plants
1859	Charles Darwin's *On the Origins of Species* is published
1865	**P.T. Westmorland joins Royal Geographical Society and studies primitive societies**
1867	**P.T. Westmorland's "The Management of Reproduction in Feeble-Minded Populations" is published**
1869	Francis Galton coins the phrase "nature versus nurture," sparks debate
1872	**P.T. Westmorland's *On the Science of Neolution* is published**
1880	**P.T. Westmorland's "Post Subjectivity and Collective Research Living" is published**
1883	Francis Galton invents the term "eugenics"
1894	**P.T. Westmorland disappears in Borneo**
1896	*The Island of Doctor Moreau* by H.G. Wells is published
1898	**P.T. Westmorland is declared dead**

1910s

c.1910	Eugenics Records Office is established at Cold Spring Harbor
	Cold River Institute is established
1914	Ella Wheeler Wilcox's *Poems of Problems* is published

1920s

| 1920 | Creutzfeldt–Jakob disease is first described |

1930s

| 1939 | Eugenics Records Office closes |

1950s

| 1951 | Henrietta Lacks dies (cervical cancer cells that were biopsied from her tumor went on to become the first human immortal cell line, |

known as HeLa cells, still used in research today)

1952	Structure of DNA is discovered
	First nuclear transfer experiments are conducted
1954	Henrietta Lacks's cells are used to produce a polio vaccine
1955	HeLa cells become the first cloned human cells
1959	**Neolution revived: "For a Better Future" (*Cambridge Pop Science Journal*)**

1960s

1961	Eisenhower's farewell address
1962	**Susan Wakefield's "Guided Genealogical Growth" is published**
	Susan Wakefield and John Mathieson meet
1963	**Susan Wakefield's "Isolating Growth Hormone in Centenarian DNA" is published**
1965	**P.T. Westmorland's return is rumored**

1970s

1970	**Dyad is established**
	Ethan Duncan and Susan Wakefield marry
1971	**Susan and Ethan Duncan publish "The Effects of Expedited Cellulose Manipulation in *Acomys cineraceus*"**
	Virginia Coady and Susan Duncan meet
	Susan Duncan and Virginia Coady conduct experiments on Yanis
1973	*Roe vs. Wade* decision
1975	First mammalian embryo is created by nuclear transfer
1976	**Project Leda begins**
1977	**Project Leda is taken over by Neolution**
	Cold River Institute is shut down
1978	First IVF baby is born
	Kendall Malone is identified for Projects Leda and Castor

1980s

1980	**Project Leda moves to DYAD**
	Project Castor recruited by military
1982	First gene patent is granted
1984	OncoMouse patent is filed
	Leda clones are born

	Castor clones are born
1985	**Virginia Coady takes over Project Castor**
1987	First description of CRISPR
1988	OncoMouse has its patent granted in the U.S., becomes first patented transgenic mammal
1989	Minnesota Twin Family Study begins

1990s

????	**Abel Johanssen is born**
1990	First gene therapy clinical trials for SCID
	Human Genome Project is launched
	The Duncans' lab in Cambridge is destroyed in a fire
	Rachel Duncan becomes first self-aware Leda clone
1991	Donna Haraway's collection of essays *Simians, Cyborgs and Women: The Reinvention of Nature* is published
	Dr. Aldous Leekie's *Neolution: The New Science of Self-Directed Evolution* is published
1993	Scientists Stillman and Hall clone human embryos
1996	Dolly the sheep is born
1997	Cumulina the cloned mouse is born

2000s

2001	**Veera Suominen discovers Leda clones undergoing experimentation in Russia**
	Helsinki incident
2002	*Harvard College v. Canada* (Commissioner of Patents): Canadian Supreme Court rejects OncoMouse patent in Canada
2003	Human Genome Project is completed
	Dolly the sheep dies of respiratory disease
	Private company Clonaid claims to have cloned a human being
2004	Assisted Human Reproduction Act is enacted in Canada, placing definitive bans and restrictions on human cloning, lab creation of chimeras, and other human experimentation
2005	**Charlotte Bowles is born**
	Kira Manning is born
	OncoMouse patent expires

2012   **Beth Childs becomes self-aware and seeks out North American Leda clones**

Brightborn begins human trials for their maggot-bots

Jennifer Fitzsimmons shows symptoms of the clone disease and seeks treatment at the Dyad Institute

Cosima Niehaus shows symptoms of the clone disease

Jennifer Fitzsimmons is first clone known to die as a direct result of the clone disease

First reported demonstration of tissue regeneration in African spiny mouse (*Acomys* sp.)

2013   *Association for Molecular Pathology v. Myriad Genetics*: Supreme Court rules that naturally occurring genes can no longer be patented, but synthetic sequences can

Scientists in Oregon successfully reprogram human somatic cells to become embryonic stem cells using SCNT; "spindle protein problem" solved

Brightborn announces clinical trials

Cosima Niehaus and Susan Duncan create a human germ line from Castor and Leda DNA

M.K. (Veera Suominen) shows symptoms of the clone disease

Rachel announces plans for new generation of clones for human experimentation

Arthur and Donald are born

Cosima and Delphine deliver the clone-disease cure to Leda clones worldwide

2015   *Science* magazine awards CRISPR "Breakthrough of the Year"

Multiple labs announce their intention to use CRISPR for human germline editing experimentation

2016   Moratorium on gene drives opposed at UN meeting

2017   First human-pig chimera embryos created

National Academies of Sciences releases guidelines on human genome editing

SELECTED
BIBLIOGRAPHY

CHAPTER ONE: "HOW MANY OF US ARE THERE?"

Campbell, K.H., et al. "Sheep cloned by nuclear transfer from a cultured cell line," *Nature* 380, no. 6569 (1996): 64–66.

Gerlach, Neil, and Sheryl N. Hamilton. "From Mad Scientist to Bad Scientist: Richard Seed as Biogovernmental Event," *Communication Theory* 15, no. 1 (2005): 78–99.

Kfoury, Charlotte. "Therapeutic cloning: promises and issues," *McGill Journal of Medicine* 10, no. 2 (2007): 112–20.

Lee, Byeong Chun, et al. "Dogs cloned from adult somatic cells," *Nature* 436, no. 641 (2005).

Machin, G. "Non-identical monozygotic twins, intermediate twin types, zygosity testing, and the non-random nature of monozygotic twinning: A review," *Am J Med Genet Part C Semin Med Genet* 151C (2009): 110–27.

Marta, M.J., et al. "A case of complete situs inversus," *Revista Portuguesa de Cardiologia* 22, no. 1 (2003): 91–104.

Matoba, S., Y. Liu, et al. "Embryonic Development following Somatic Cell Nuclear Transfer Impeded by Persisting Histone Methylation," *Cell* 159, no. 4 (2014): 884–95.

Moorman, Antoon, et al. "Development of the heart: (1) formation of the cardiac chambers and arterial trunks," *Heart* 89, no. 7 (2003): 806–14.

Peat, J., and W. Reik. "Incomplete methylation reprogramming in SCNT embryos," *Nature Genetics* 44 (2012): 965–66.

Qui, Sara D., Paul D. Smith, and Peter F. Choong. "Nuclear reprogramming and induced pluripotent stem cells: a review for surgeons," *ANZ Journal of Surgery* 84, no. 6 (2014): E1–E11.

Shiels, Paul G., et al. "Analysis of Telomere Length in Dolly, a Sheep Derived by Nuclear Transfer," *Cloning* 1, no. 2 (2004): 119–25.

Tachibana, Masahito, et al. "Human Embryonic Stem Cells Derived by Somatic Cell Nuclear Transfer," *Cell* 153, no. 6 (2013): 1228–38.

Wakayama, S., et al. "Successful serial recloning in the mouse over multiple generations," *Cell Stem Cell* 12, no. 3 (2013): 293–97.

Wakayama, T., et al. "Full-term development of mice from enucleated oocytes injected with cumulus cell nuclei," *Nature* 394 (1998): 369–74.

CHAPTER TWO: "THERE'S ONLY ONE OF ME"

Hughes, C., et al. "Origins of individual differences in theory of mind: From nature to nurture?" *Child Development* 76 (2006): 356–70.

Jang, Kerry L., and W. John Livesley. "Heritability of the Big Five Personality Dimensions and Their Facets: A Twin Study," *Journal of Personality* 64, no. 3 (1996): 577–92.

Kendler, Kenneth S., et al. "A Longitudinal Twin Study of Personality and Major Depression in Women," *Archives of General Psychiatry* 50, no. 11 (1993): 853–62.

Okbay, A., et al. "Genetic variants associated with subjective well-being, depressive symptoms, and neuroticism identified through genome-wide analyses," *Nature Genetics* 48, no. 6 (June 2016): 624–33.

Olson, J. M., et al. "The heritability of attitudes: A study of twins," *Journal of Personality and Social Psychology* 80 (2001): 845–60.

CHAPTER THREE: "YOU'RE JUST A BAD COPY OF ME"

Brenner, J. Chad. "Genotyping of 73 UM-SCC head and neck squamous cell carcinoma cell lines," *Head & Neck* 32, no. 4 (2010): 417–26.

Fehilly, Carole B., S.M. Willadsen, and Elizabeth M. Tucker. "Interspecific chimaerism between sheep and goat," *Nature* 307 (1984): 634–36.

Miech, Ralph P. "The role of fetal microchimerism in autoimmune disease," *International Journal of Clinical and Experimental Medicine* 3, no. 2 (2010): 164–68.

Van Dijk, Bob A., Dorret I. Boomsma, and Achile J.M. de Man. "Blood Group Chimerism in Human Multiple Births Is Not Rare," *American Journal of Medical Genetics* 61 (1996): 264–68.

Yu, Neng, et al. "Disputed Maternity Leading to Identification of Tetragametic Chimerism," *The New England Journal of Medicine* 346 (2002): 1545–52.

Yunis, Edmond J., et al. "Chimerism and tetragametic chimerism in humans: implications in autoimmunity, allorecognition and tolerance," *Journal of Immunology Research* 38 (2007): 213–36.

CHAPTER FOUR: "THIS IS MY BIOLOGY, MY DECISION"

"About Human Germline Gene Editing," Center for Genetics and Society, 2016. https://www.geneticsandsociety.org/internal-content/about-human-germline-gene-editing.

Bézier, Annie, et al. "Polydnaviruses of Braconid Wasps Derive from an Ancestral Nudivirus," *Science* 323, no. 5916 (2009): 926–30.

"Pterygium Popliteal Syndrome," Genetics Home Reference. NIH U.S. National Library of Medicine, 2008. https://ghr.nlm.nih.gov/condition/popliteal-pterygium-syndrome.

Kaur, Ishwinder. "Interleukin-4-Inducing Principle from Schistosoma mansoni Eggs Contains a Functional C-Terminal Nuclear Localization Signal Necessary for Nuclear Translocation in Mammalian Cells but Not for Its Uptake," *Infection and Immunity* 79, no. 4 (2011): 1779–1788.

Lanphier, Edward, et al. "Don't edit the human germ line," *Nature* 519 (2015): 410–11.

Packer, Michael S., and David R. Liu. "Methods for the directed evolution of proteins," *Nature Reviews Genetics* 16 (2015): 379–94.

CHAPTER FIVE: "MY POOR, POOR RACHEL"

Alvarez, Julie A., and Eugene Emory. "Executive Function and the Frontal Lobes," *Neuropsychological Review* 16, no. 1 (2006): 17–42.

Devinsky, Orrin. "Executive Function and the Frontal Lobes," in *Neurology of*

Cognitive and Behavioral Disorders. Oxford: Oxford University Press, 2004: 302–29.

Faraz Kazim, Syed, et al. "Management of penetrating brain injury," Journal of Emergencies, Trauma, and Shock 4, no. 3 (2011): 395–402.

Gordon, D.S. "Penetrating Head Injuries," Ulster Medical Journal 57, no. 1 (1988): 1–10.

Harmon, Katherine. "The Chances of Recovering from Brain Trauma: Past cases show why millimeters matter," Scientific American. http://www.scientific american.com/article.cfm?id=recovering-from-brain-trauma.

Kotowicz, Zbigniew. "The Strange Case of Phineas Gage," History of the Human Sciences 20, no. 1 (2007): 115–31.

Lundgren, Kristine, Nancy Helm-Estabrooks, and Reva Klein. "Stuttering Following Acquired Brain Damage: A Review of the Literature," J Neurolinguistics 23, no. 5 (2010): 447–54.

MacMillan, Malcolm, and Matthew L. Lena. "Rehabilitating Phineas Gage," Neuropsychological Rehabilitation 20, no. 5 (2010): 641–58.

Stone, James L. "Transcranial Brain Injuries Caused by Metal Rods or Pipes Over the Past 150 Years," Journal of the History of the Neurosciences 8, no. 3 (1999): 227–34.

Wilgus, Jack, and Beverly Wilgus. "Face to Face with Phineas Gage," Journal of the History of the Neurosciences 18, no. 3 (2009): 340–45.

CHAPTER SIX: "YOUR LITTLE GIRLS ARE DYING"

Gore, M.E. "Adverse effects of gene therapy: Gene therapy can cause leukaemia: no shock, mild horror but a probe," Gene Therapy 10 (2003): 4.

Hillier, Stephen G., and Griff T. Ross. "Effects of Exogenous Testosterone on Ovarian Weight, Follicular Morphology and Intraovarian Progesterone Concentration in Estrogen-Primed Hypophysectomized Immature Female," Biology of Reproduction 20 (1979): 261–68.

Shammas, Masood A. "Telomeres, lifestyle, cancer, and aging," Current Opinion in Clinical Nutrition & Metabolic Care 14, no. 1 (2011): 28–34.

Westergard, Laura, Heather M. Christensen, and David A. Harris. "The Cellular Prion Protein (PrPC): Its Physiological Function and Role in Disease," Biochim Biophys Acta. 1772, no. 6 (2007): 629–44.

CHAPTER SEVEN: "AND WE, HERE, SHALL DRINK FROM THE FOUNTAIN FIRST"

Atzmon, Gil, ed. *Longevity Genes: A Blueprint for Aging.* New York: Springer-Verlag, 2016.

Darwin, Charles. *The Voyage of the Beagle.* 1839.

Kragl, Martin, et al. "Cells keep a memory of their tissue origin during axolotl limb regeneration," *Nature* 460 (2009): 60–65.

Urbach, Achia, et al. "Lin28 sustains early renal progenitors and induces Wilms tumor," *Genes Dev.* 28, no. 9 (2014): 971–82.

"World Health Statistics 2016: Monitoring Health for the SDGs," World Health Organization, 2016. http://www.who.int/gho/publications/world_health_statistics/2016/en/.

ACKNOWLEDGMENTS

SARAH

I can't do this without you, Cosima.

(2.10 "By Means Which Have Never Yet Been Tried")

Back in 2014, BBCAmerica held its first *Orphan Black* fan meetup in San Diego, California. We were strangers, but Casey won the event's trivia contest and Nina was one of the winners of the theater challenge, and we got to share the stage with the cast and creators of the show. We didn't talk to each other at all — there was too much excitement! — but Clone Club reconnected us online, like some sort of fandom missed connection.

There are so many people who have helped to shape this book, starting with that day in 2014. Thank you to Sam Maggs for taking us on as *Orphan Black* science recap writers for *The Mary Sue*, and to the editors who have worked with our recaps since: Jess Lachenal, Carolyn Cox, and Sam Riedel. Thank you to Maria Vicente for seeing the seeds for a book in our science recaps and for being an all-around rockstar agent. Thank you to our editor, Crissy Calhoun, for her enthusiasm and humor (and for keeping us in check whenever we spiraled excitedly into science jargon).

Thank you to the entire *Orphan Black* team: John Fawcett and Graeme Manson, the cast and crew, David Fortier, Ivan Schneeberg and everyone at Temple Street, for creating this amazing clone show, and special thanks to Cosima Herter.

And, of course, thank you, Clone Club.

CASEY: I would like to thank the following awesome Clone Clubbers who shared in my love for *Orphan Black*, added fuel to my science fire, and have left me with lasting friendships: Johanna Vique, Kim Hawker, Stevie Militello, Theressa Goldberg, Emery Lu, Charlie Beckler, Dawson Schachter, Jacqueline Bircher, Milena Battaglia, Hafsah Mijinyawa, and Irina Heimerle. A special thanks to all my friends and family who encouraged my love for *Orphan Black* and the channeling of that love into this book. Thanks to my parents and sister for putting up with my obsession, for attempting to watch the show even though they couldn't keep the clones straight, and for supporting me throughout my life. And thank you to my husband, Eric, for being with me every step of the way with this book, from the meetup in 2014 until today, and every day for the last eight years.

NINA: I'd also like to thank the following who generously and unquestioningly shared their science or *Orphan Black* expertise: Dan Chaput, Dana Falcioni, Natalie Eisen, and Josh Osika. Thank you to Cosima Herter, Matt Wakefield, and Tatiana Maslany for patiently answering my questions. Thanks to my friends in Clone Club, IRL, and my girl gang (and all three sometimes Venn) who cheered me on and who put up with me constantly finding ways to talk about *Orphan Black*; to my parents and family for supporting me even though they never quite understood my passion for the show. And thank you to my wife, Cora, for being my constant every step of the way.

Published by ECW Press
665 Gerrard Street East
Toronto, Ontario, Canada M4M 1Y2
416-694-3348 | info@ecwpress.com

Purchase the print edition and receive the eBook free!
FOR DETAILS, GO TO ECWPRESS.COM/EBOOK

LIBRARY AND ARCHIVES CANADA
CATALOGUING IN PUBLICATION

Griffin, Casey (Scientist), author
The science of Orphan black : the official
companion / Casey Griffin and Nina Nesseth ;
with Graeme Manson and Cosima Herter.

Issued in print and electronic formats.
ISBN 978-1-77041-380-1 (paperback)
also issued as: 978-1-77305-045-4 (pdf);
978-1-77305-044-7 (epub)

1. Orphan black (Television program).
2. Science in mass media. 3. Genetics in mass
media. 4. Bioethics on television. 5. Human cloning
in mass media. I. Nesseth, Nina, author II. Manson,
Graeme (Writer), author III. Herter, Cosima, author
IV. Title.

PN1992.77.075G75 2017   791.45'72
C2016-906420-4   C2016-906421-2

Editor for the press: Crissy Calhoun
Cover and text design: Michel Vrana

TEMPLE STREET   BOAT ROCKER BRANDS   ECW PRESS

The publication of *The Science of Orphan Black* has been generously supported by the Government of
Canada through the Canada Book Fund. *Ce livre est financé en partie par le gouvernement du Canada.*
We also acknowledge the contribution of the Government of Ontario through the Ontario Book Publishing
Tax Credit and the Ontario Media Development Corporation.

PRINTED AND BOUND IN CANADA   PRINTING: FRIESENS   5   4   3   2   1